HOW SCIENTISTS FIND OUT

about

matter time space energy

Herman Schneider

illustrated with line drawings and photographs

McGraw-Hill Book Company

New York St. Louis San Francisco Auckland Düsseldorf Johannesburg
Kuala Lumpur London Mexico Montreal New Delhi Panama Paris
São Paulo Singapore Sydney Tokyo Toronto

Library of Congress Cataloging in Publication Data
Schneider, Herman, date.
How scientists find out.

Bibliography: p.
Includes index.
SUMMARY: Discusses the scientific investigations of matter, time, space, and energy with suggested experiments for further exploration.
1. Science—Juvenile literature. [1. Science]
I. Title.
Q163.S4523 501'.8 76-18142
ISBN 0-07-055447-1
ISBN 0-07-055448-X lib. bdg.

Book Design and illustrations by Melinda Wooster, A Good Thing, Inc.

123456BPBP789876

With gratitude, for many valuable suggestions, to the author's friends and colleagues:

Mr. Henry Berman, former Director of the Scientific Instrument Division, Carl Zeiss-Jena, U.S.A.

Dr. Robert L. Chase, Senior Electronics Engineer, Brookhaven National Laboratory, Upton, N.Y.

Mr. Max Eisner, former physics teacher, Fort Hamilton High School, New York.

Mr. Leo Schneider, Senior Science Editor, The New Book of Knowledge Encyclopedia.

Dr. Edward A. Spiegel, Professor of Astronomy, Columbia University, New York.

contents

introduction

HOW DO WE KNOW?

This moon pebble is three billion years old.

This moon pebble is a billion years older.

How Do We Know?

What's in the pebbles that reveals their age?

A drop of blood contains over a thousand different substances. Each one can be separated from the rest and identified.

How Do We Know?

How can we take apart a drop of blood and identify all those substances?

The surface temperature of the planet Venus is over 900°F. Its atmosphere contains large amounts of carbon monoxide and sulfuric acid.

How Do We Know?

How do we take the temperature of a planet millions of miles away? How do we examine and label the chemicals in its atmosphere?

Of course you know the answer to that repeated question, "How do we know?" We know because scientists tell us so.

But How Do Scientists Know?

What instruments do they use, what methods do they work by, to explore the tiny and the huge, the far away and long ago?

This book will help you to discover how scientists find out. You will also find directions for making simple scientific instruments out of materials in your kitchen and your toolbox. That's another way of finding out How Scientists Find Out.

What Is a Scientific Instrument?

You can see seven stars in the Big Dipper—if you're looking with the naked eye. With a small telescope you can see about twenty more in the same space. With a large telescope you can see about two thousand!

Telescopes are scientific instruments for *extending a human sense*—in this case, the sense of sight.

Right now, the earth's magnetic field is sweeping through you, but you don't feel a thing. Walk near an iron mine, where the magnetic field swings toward the mine, and you still don't feel anything. But a magnetic compass does; its needle turns and lines up with the changed direction of the field. That's why a geologist carries a magnetic compass when exploring and prospecting.

A magnetic compass is a scientific instrument that *provides a nonhuman sense*—in this case, a sense of magnetism.

A test tube full of seawater may contain a million tiny plants and animals. A marine biologist could spend days looking for them, one drop of water at a time. But that isn't necessary. Put the tube in a centrifuge; it spins at high speed and forces the solid particles—the plants and animals—out to the end of the tube, where they can be picked out in a cluster.

A centrifuge is a scientific instrument for *processing*: it performs a separation process on the seawater.

A doctor connects a set of wires to a patient's skin, using suction cups to make contact. The wires come from a machine called an *electrocardiograph.* Each beat of the heart sends electrical impulses that are recorded as a line on a graph. The height of the line shows the strength of the heartbeat. The space from one peak to the next shows the duration of each beat.

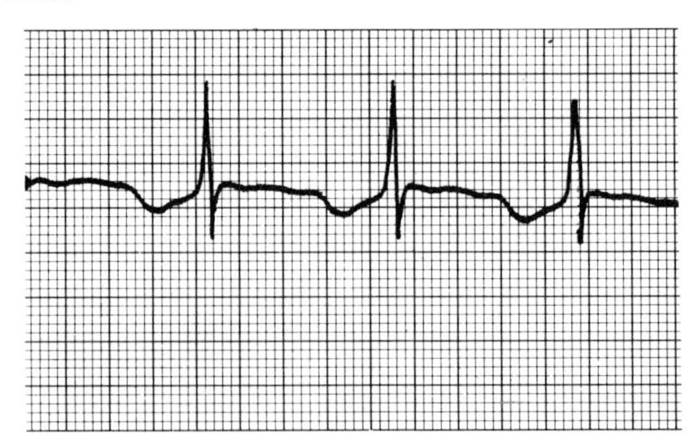

An electrocardiograph is a scientific instrument that *measures, counts, and records* heartbeats.

These, then, are the main functions of most scientific instruments:

- to extend a human sense, or
- to provide a nonhuman sense, or
- to act as a processor, or
- to measure, count, and record, or
- to do a combination of these jobs.

You don't have to visit a scientific laboratory to find instruments that perform these functions. You have many of them in your home. Look at the following list. You may like to classify the items according to the functions they perform.

thermometer	measuring cup
binoculars	egg timer
air conditioner	barometer
thermostat	humidity meter
electric meter	clock
tape recorder	eggbeater
tea strainer	stove
tape measure	gas meter
magnifying glass	light meter
electric fan	coffee filter

If you want to classify these items, be sure to give each one all the credit it's entitled to. For instance, a thermometer is entitled to credit for *two* functions: (1) It *extends* your sense of touch. You don't have to dip your finger in the bathwater and report "Ouch, it's too hot." (2) It gives a more accurate report than your finger, because it can *measure* the hotness in degrees.

A thermostat is entitled to three credits: it's a thermometer (two credits) *and* it acts as a *processor*: It turns on the heating system when the room temperature drops below a certain point.

Now let's find out about some instruments that scientists use, and how they use them as they explore the framework of science: Matter, Time, Space and Energy.

1 matter

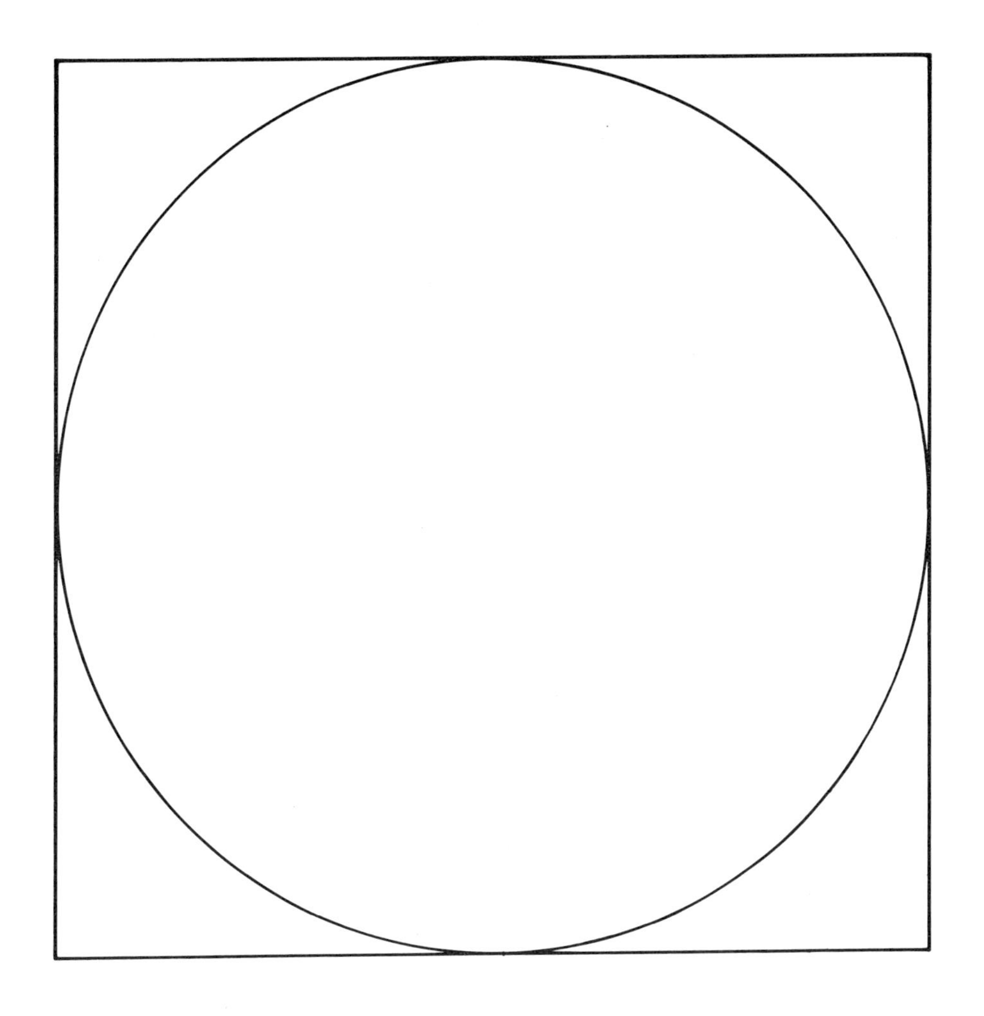

"What's in it?"

Scientists often ask that question. The "it" might be a piece of moon rock, or vapor puffing out of a newly erupted volcano, or a drop of blood from a diseased person.

Usually the "it" is really "them," because most substances are *mixtures.* Moon rocks are mainly mixtures of minerals such as feldspar, quartz and calcite. A volcano's vapors contain gases such as carbon dioxide, hydrogen sulfide, and water vapor. Blood contains more than a thousand different gases, liquids and solids.

Separators

The first step in finding out "what's in it" is to separate the mixture into its different parts. This is called *analysis.* We analyze a mixture into its different parts by taking advantage of the differences. It's what a cook does by pouring a potful of boiling spaghetti into a strainer. The spaghetti is too large to pass through the holes in the strainer, while the water easily passes through. Some coffeemakers contain filter paper to separate coffee grounds from liquid coffee.

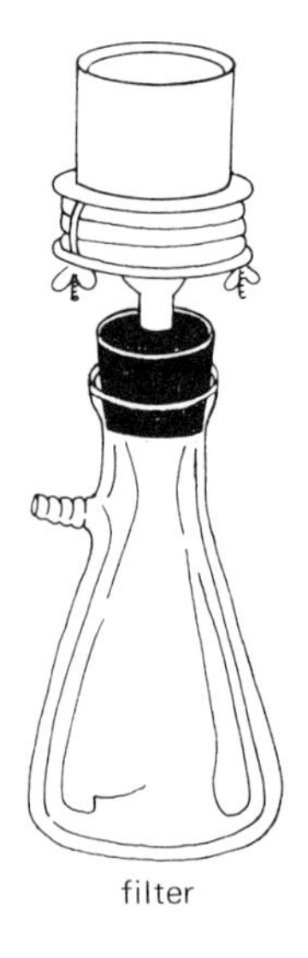
filter

Analysis By Size

Scientists, too, use strainers and filter papers to separate solids from liquids. But most separation jobs are not that easy. Bacteria, for instance, can pass easily through the holes in paper, so there's no use trying to trap them that way. But filters made of clay have holes small enough to do the job.

Maybe you've noticed how the outside of a clay flowerpot becomes damp when you water the soil inside. The water can pass through the clay, but the soil and soil bacteria are held back. Scientists use filters made of fine white clay, called *unglazed porcelain,* to filter out bacteria and other small particles from liquids.

Analyzing By Weight

Another way of separating solids from liquids is by allowing gravity to do the job. When somebody makes a muddy mess out of a swimming pond, the problem can be solved by not doing anything. The muddy stuff is heavier than water, so it settles to the bottom in an hour or so—or a day, or longer.

But nobody wants to wait around that long if there's a faster way. If gravity were stronger, the particles would settle down faster. We can't change the force of gravity, but in a laboratory we can use another force in its place. That force is called *centrifugal force.* It's produced by spinning, in a machine called a *centrifuge.* The heaviest particles are forced out first, then the next heavier, and so on. It does the same job much faster than gravity.

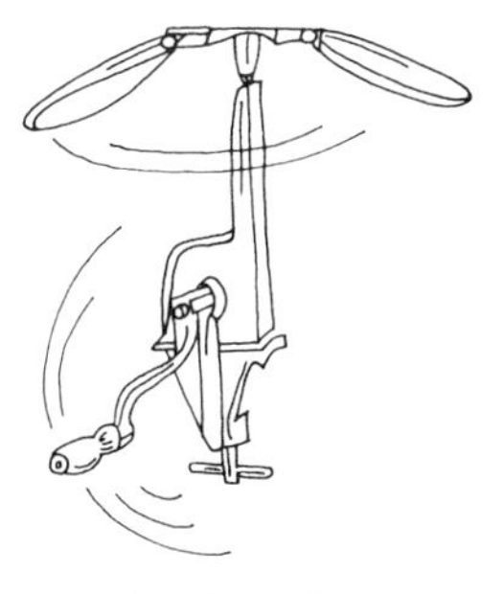

hand centrifuge

For example, a doctor wants to test a sample of blood for a substance called *cholesterol.* This is present in the liquid part of the blood, the serum. To get an accurate test, the serum has to be separated from the nonliquid part, the red cells, white cells, and other solids. The blood is centrifuged for fifteen minutes. The different substances are separated into layers, according to their weight.

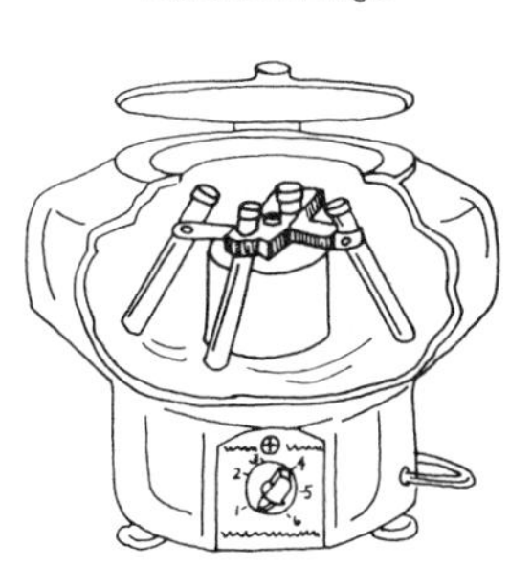

electric centrifuge

So far, the job of separating materials hasn't been too difficult. Solids are clearly different from liquids, and we take advantage of the differences. We use strainers, paper filters, clay filters, gravity or centrifugal force to separate solids from liquids, or heavy solids from light solids.

Now we come to a more difficult job: to separate liquids from liquids. What differences can we take advantage of? Let's take a hint from the cook in the kitchen.

Analyzing By Heat

Clear beef soup is a mixture of liquids: water, beef juice, and others. If the soup is too weak and watery we can get rid of some of the water. By just simmering gently, the water gradually boils away but the beef juice doesn't. The soup becomes thicker by losing some water.

Where is the difference that we took advantage of? It's in the boiling point. Under ordinary conditions, water boils at 100°C. (212°F.). The boiling point of beef juice is several degrees higher. If we heat the soup to a temperature between these two boiling points, we separate one liquid from the other. We get rid of some of the water, and retain the beef juice.

Scientists don't usually want to get rid of anything, especially when trying to determine what it is. So they catch the boiled-away material and cool it in another container. It condenses back into a liquid, and can be examined separately.

Here's a way to see this for yourself. Put a few drops of food coloring or coffee in a pan full of water. Heat it to a gentle boil. Catch the vapor against a cool pan lid. Let the drops fall into a dish. Is there color in the drops?

Try it again with sugar and water. Is there sweetness in the drops?

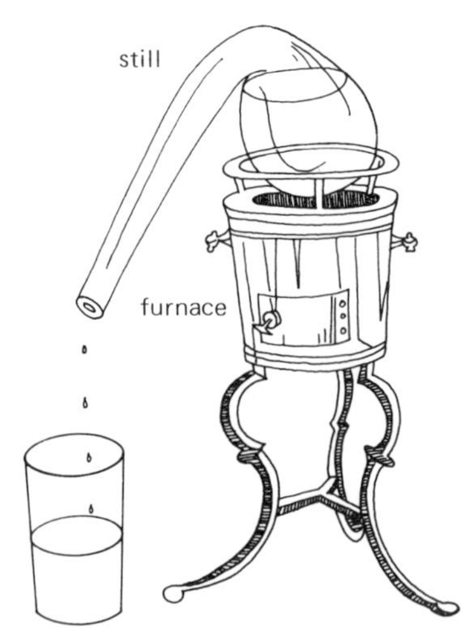

This process of heating and cooling to separate a mixture of liquids is called *distillation.* The apparatus for distillation is called a *still.*

Here's one of the earliest types of still, called a *retort.* A flame heats the mixture of two liquids in the bowl of the retort. The liquid with the lower boiling point turns into a vapor and passes down through the tube. On the way the vapor cools and condenses into drops of liquid. A container catches the liquid, a drop at a time.

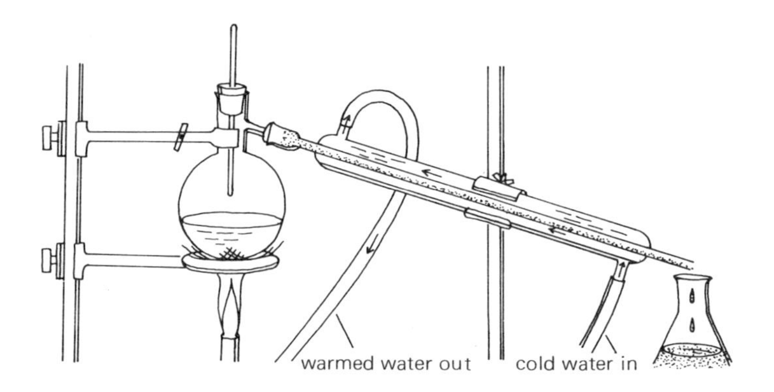

A modern still works the same as a retort, but with an improvement. Cold water flows around the long tube through which the vapor passes. This causes the vapor to condense more quickly than in a retort.

We don't have to limit ourselves to mixtures of two liquids. We can distill mixtures of any number of liquids, so long as they have different boiling points. For example, crude oil (petroleum) is a mixture of liquids of different boiling points. These liquids can be distilled into gasoline, kerosene, lubricating oil, diesel oil, furnace oil, and hundreds of other substances. (Even what is left over, the solid material, becomes useful asphalt and tar.)

The crude oil from different wells contains different proportions of gasoline, kerosene and other substances. So when oil is first struck, a sample is sent to the laboratory to be distilled and analyzed. The bulk crude oil is distilled in huge stills called *oil refineries*, like the one shown on the next page.

It looks enormous and complicated, because it is. Yet it consists mainly of three basic kinds of parts: (1) parts where crude oil is heated into its separate vapors; (2) parts where the separate vapors are cooled and condensed into separate liquids; (3) pumping machines

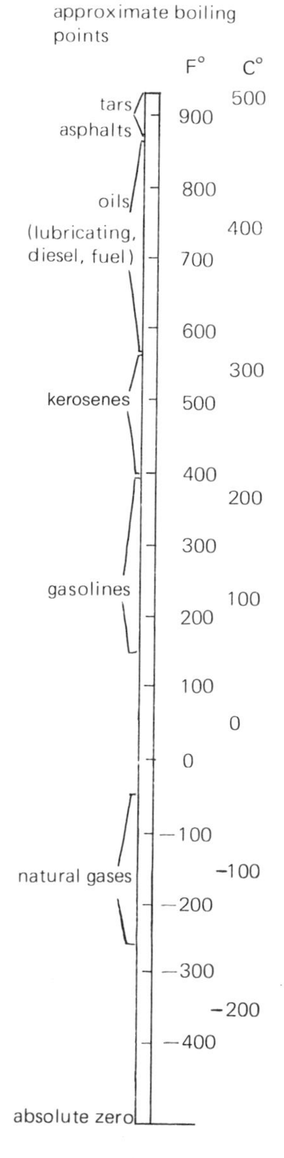

that move liquids and vapors through pipes from part to part.

Analysis By Cooling And Warming

Pure air is a mixture of several invisible gases: oxygen, nitrogen, and several others. But how do we know this if pure air is invisible? We can't distill air into its separate gases because it's already in the gas state. However, if we cool air enough, it becomes a liquid. Then we can distill it and catch the different gases as they boil off at different temperatures. And we can use the gases.

In air-liquefaction factories air is cooled and liquefied by machines that work like your refrigerator. These machines, however, can bring the temperature down to about 270°C. *below zero.* (The low point in a home freezer is about 20 below.) At −250°C°., air is a clear liquid that looks and pours like water—but don't dip your finger in it!

The liquid air is warmed up in stages, inside a system of pipes. The first substance to turn into a gas is helium. Then comes neon. This is piped off and stored in tanks, to be used in making electric signs that glow with a red-orange color. Then comes nitrogen, used in some spray cans and in making foam plastics such as styrofoam. Next is argon, used in ordinary electric bulbs. Then comes oxygen, which has many uses, from helping sick people breathe to welding and cutting steel. Krypton and xenon are used in certain kinds of electric lamps. (Very small amounts of other gases are also present.)

Analysis By Magnetism

Sweep a magnet through a basketful of trash. The magnet will pick out nails, tacks, bottle caps, paper clips, and other articles made of *ferrous metals* (iron and steel). The magnet will leave behind pencil shavings, copper pennies, rubber bands, shredded love letters and other nonferrous materials.

This property of magnets, to separate ferrous from nonferrous, is used by scientists and engineers. A mining engineer, for example, can analyze a sample of crushed iron ore. He can determine whether there's enough iron in it worth mining.

This is a diagram of a continuous magnetic analyzer. The magnets inside the rotating drum cause ferrous materials to cling for a little way. The nonferrous materials drop off first.

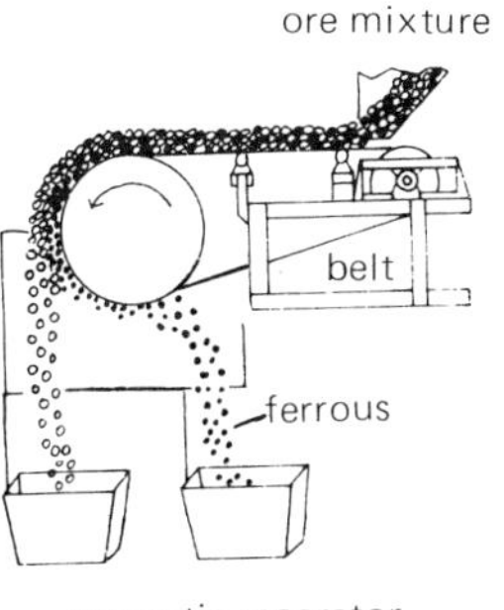

magnetic separator

Coin-operated slot machines contain magnets. These attract counterfeit coins and discs made of iron and steel. They allow good coins, made of copper, silver, and other nonferrous metals to move on and operate the machine.

Analysis By Absorption

When you dip the corner of a sugar cube into a cup of coffee, you can see the coffee run up the cube, absorbed by the sugar. If you organized a race between tea and coffee, you would see one of the liquids climb the sugar

cube faster and higher than the other. (The tea, if both are equally concentrated.)

Every liquid moves at its own special rate through sugar cubes, through sand, even through paper. This property makes it easy to analyze a mixture of two or more liquids. Just send the mixture through an absorber and wait for results.

chromatography

beginning of experiment

later (enlarged)

Try it yourself. Mix several drops of different colored inks or food colorings. Put a couple of drops of the mixture near one end of a strip of white (but not shiny) paper. Or draw a scramble of lines on the paper using different colored felt tip pens.

Place the strip of paper in a cup or tumbler. Add water to just below the colored mixture. Watch the water rise through the paper and lift the coloring matter. Each color is lifted at its own rate because its materials are different from the other colors. Watch the mixture become analyzed into separate bands of color!

The Greek word for "color" is *khroma*. This system of analysis with paper is called *paper chromatography*.

Let's consider paper chromatography for a moment. The paper can't see colors. So the liquids are probably being separated not by color, but by some other property such as stickiness, or weight, or thinness. Therefore we should be able to separate different liquids of the same color, or of no color. For example, a mixture of starch and sugar is white. Neither substance leaves a color on paper. But we can test for each substance. (For example, a drop of brown iodine produces a purple stain with starch, and a brown stain with sugar. For more tests, see the section on reagents, page 18.)

Analysis By Electrochromatography

Perhaps you are not familiar with that word, but you can guess its meaning. It's chromatography performed with the aid of an electric current. This very simplified diagram shows the main parts of the apparatus.

A tube (A) contains the liquid mixture to be analyzed (for example blood *serum*, the liquid part of blood). The liquid slowly oozes onto a sheet of wet paper (B). The paper is wetted with a liquid that can conduct electricity. On each side of the paper is a metal strip (C). The two strips are attached by wires to a source of electric current. This is shown by a plus sign (+) and a minus sign (−). The current flows sideways across the paper.

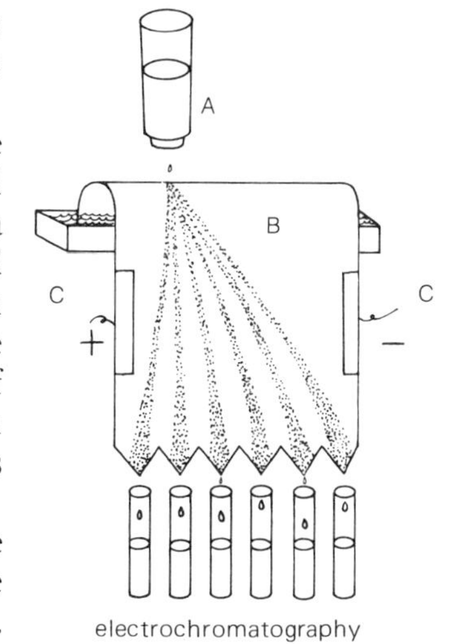

electrochromatography

As the current flows across, it meets the mixture of blood serum oozing down. The moving current pushes the liquids sideways. But each part of the mixture is different from the other parts, and is carried a different distance. (Just as in simple paper chromatography.)

So now we have several streams of liquids flowing down and across. Each stream is a separate part of the mixture. When the streams reach the bottom, they drip off the pointed ends of the paper into separate test tubes. The blood serum has been separated into its different portions, or *fractions*.

Analyzing By Smashing

Can you blow on a hammer, suspended from a string, to break a glass jar?

You can if you blow it right. You blow a little puff of air to start the hammer swinging forward in a very tiny arc. As it swings back you give it another puff, adding energy to increase the arc a tiny bit. Puff by puff, at the end of each backward swing, you continue to add energy and increase the arc. With enough puffs you can blow the hammer all the way to the jar. More puffs, more energy, and you can smash the glass jar, which is full of gold coins.

A little energy at a time, if it comes at the right time, can go a long way. That's the idea behind the jar-smashing system. It's also the idea behind a group of scientific instruments called cyclotrons, betatrons, synchrotrons, and linear accelerators. The common name for the whole group is *atom smashers.* As with most common names, it isn't quite accurate. It should be *nucleus smashers.*

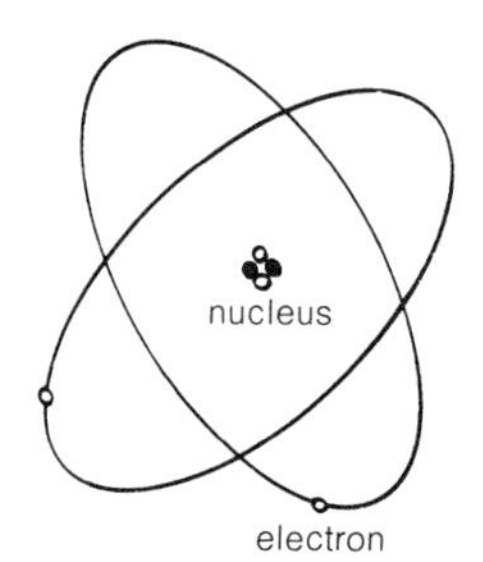

Let's review a bit.

ATOM. The smallest particle of a single kind of substance, or element. For example, the smallest particle of the element carbon is one atom of carbon. An atom consists of a central part called the *nucleus,* surrounded by whirling, spinning particles called *electrons.* Right now we're interested in the nucleus.

NUCLEUS. There are all kinds of nuclei. The smallest and lightest is the nucleus of ordinary hydrogen. This consists of one single particle called a *proton,* and nothing else. The nuclei of other elements are larger and heavier because they contain more protons plus other particles called *neutrons.*

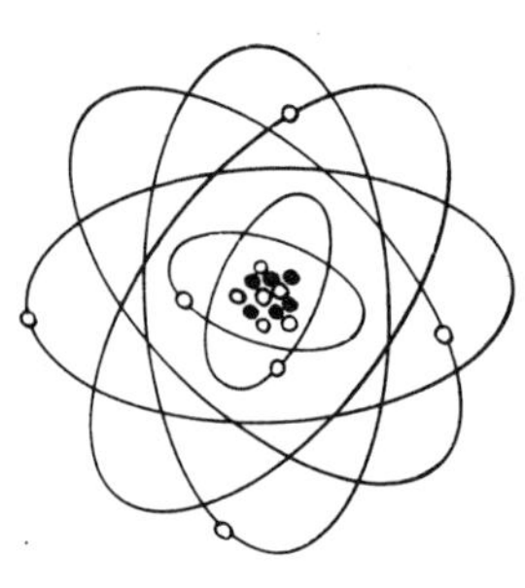
atom of Carbon

All protons are alike. All neutrons are alike. All electrons are alike.

It is the *number* of protons and electrons that makes the difference between an atom of lead and an atom of gold, between iron and oxygen.

Can we take apart atoms of lead and rebuild them into atoms of gold? Or can we even just take apart different kinds of nuclei to study them better? It isn't easy because of—

The Strong Nuclear Force

The force that holds protons and neutrons together in a nucleus is unbelievably powerful. It's much much more powerful than gravity. Here's an example of how strong it is:

Drop a pin from a height of one foot above your desk. The force of gravity pulls it down to your desk with a faint tap. If gravity were as strong as the nuclear force, that pin would plunge right through the desk, through the floor, through the soil and rock, down, down to the center of the earth!

A fantastic force indeed, holding together those nuclear particles! The nuclear force is too strong to allow us to split the nucleus in ordinary ways. So let's try an extraordinary way, the blowing-the-hammer way of splitting nuclei.

To hammer things as tiny as nuclei, we need tiny hammers. Electrons or protons are tiny. Instead of swinging them back and forth, as with the hammer, they are whirled around and around by a powerful electric current in a machine called a *cyclotron.* Faster and faster they whirl. As they gain speed they whirl in larger and larger circles (like the larger arcs of the hammer) until they smack into the target substance (like the jar) and split it apart.

All kinds of things come tumbling from the nuclei—protons and neutrons of course, as expected—but many unexpected things as well. There are particles called muons, pions,

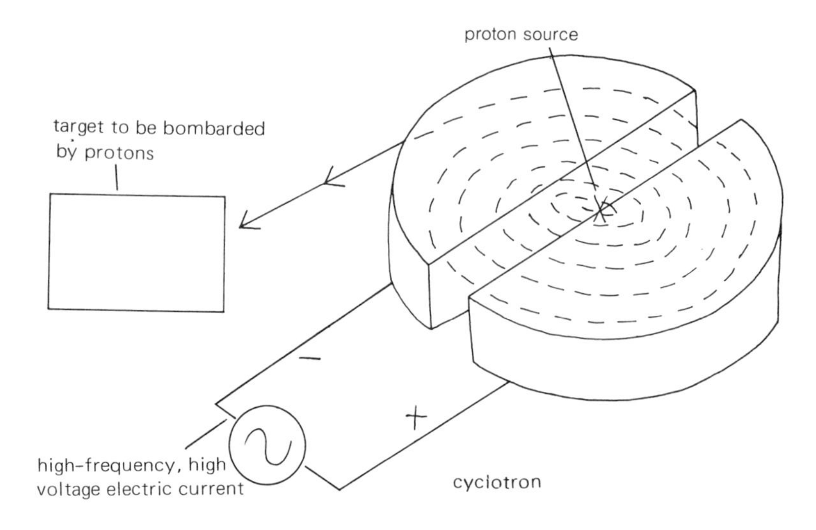

positrons, neutrinos, photons, and many other strangely named nuclear bits.

What are they? Where did they appear from? Why do they vanish in a billionth of a second? Can we make use of them in any way?

All kinds of questions are being explored by scientists, at nucleus smashers all over the world.

Reagents

Gold! Gold! Gold! The prospector jumps for joy.

Or *is* it gold? Those glittery yellow sparkles in the sand could be real gold—worth about $2,500 a pound—or they could be fool's gold, iron pyrites—ten cents a pound. So, before the prospector hurries off to celebrate at the Klondike Saloon, he needs to make sure he has something to celebrate about. He has to test for gold.

He rushes back to town, to the assay office, with a sample of the glittery stuff. An assayer will test it.

A substance for testing the identity of something is called a *reagent.* Scientists have thousands of reagents. Some are used for testing single substances; some are used for testing groups of substances. Let's look at a few examples of each.

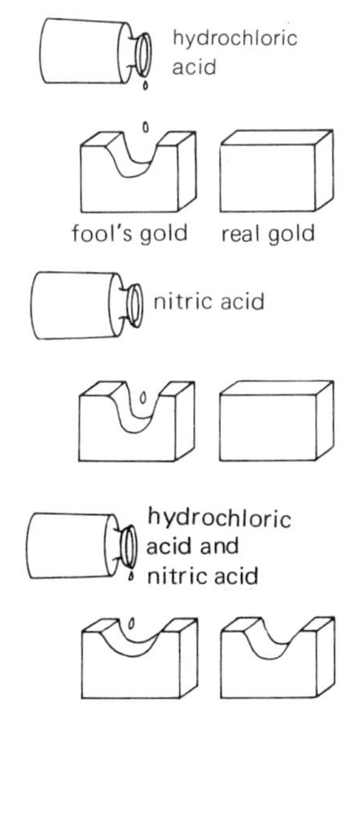

GROUP TESTS. You have been using the idea of group testing for a long time. For example, when you see a strange animal on the window-sill, you don't ask yourself, "Is that a new kind of grasshopper, or a duck, or a frog?" You see a beak, two legs and feathers, and you know it's a bird. You have grouped the new animal. You have eliminated the many millions of things it is *not.*

You still have to find out what kind of bird it is, but your work is much easier than if you had to go through the whole animal kingdom.

Scientists don't have such an easy time as you, because most substances are not equipped with beaks and feathers. You can *look* at a white powder and *guess* that it's sugar, or salt, or starch. But you wouldn't (and shouldn't) taste it. It might be another white powder, arsenic trioxide, a deadly poison. So we use reagents to test substances or groups of substances.

TESTING FOR ACIDS. One large group of substances is the *acid* group. All acids have a sour taste. (But remember why we don't test by tasting!) There are thousands of acids, but all of them can be tested with the same reagent, called *blue litmus.* This comes in little paper strips, which you can buy in any drugstore. A quarter's worth will put you in business.

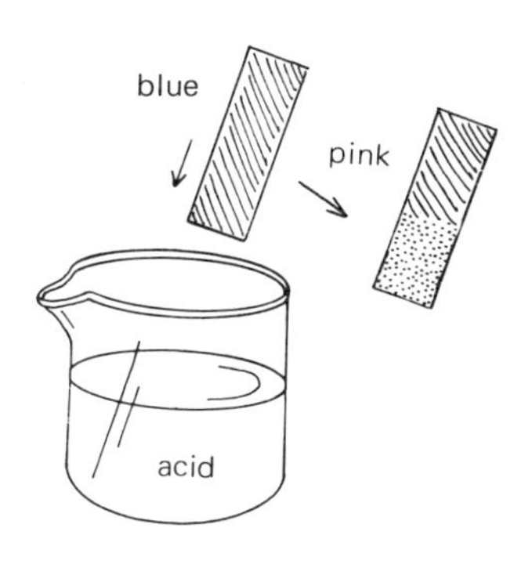

Lemon juice belongs to the acid group. If you dip the end of a strip of blue litmus paper in lemon juice, the paper will turn pink. All acids do this to blue litmus paper. No non-acids do this. So blue litmus is a group test for acids.

You might like to test the following, to see

which are acids: vinegar, soda water, ammonia water, salad oil, boric acid, sour milk, fresh milk, yogurt, milk of magnesia, grapefruit juice, orange juice, saltwater, ink, watercolor paint, maple syrup, bicarbonate of soda.

Maybe you've noticed that lemon juice causes tea to turn pale. Is it because the lemon juice is pale and mixes with the dark tea? Try two cups of tea, of the same darkness. Add a teaspoon of lemon juice to one. Add a teaspoon of water to the other. Then compare the color in each cup.

Could tea be used as a reagent for testing lemon juice? If so, could it also be used for testing other acids? Try a few of the substances you found to be acids when you tested with blue litmus paper.

Another homemade test for acids is done with the juice from boiled red cabbage. Let the juice stand for an hour after boiling the cabbage, until it turns purple. Acids will turn it red.

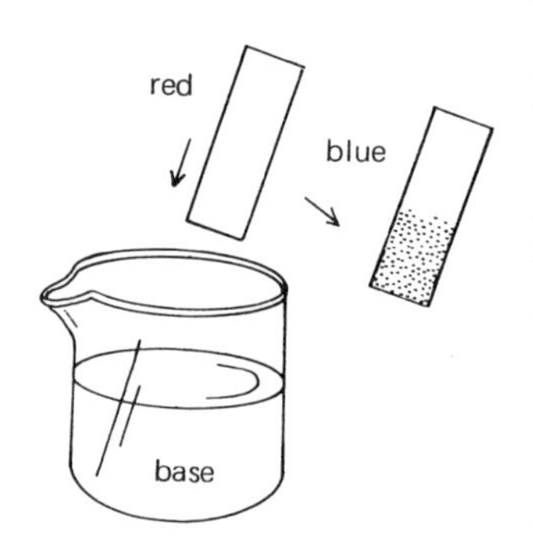

A GROUP TEST FOR BASES. There's another large group of substances called *bases,* or *alkalis.* These can be tested with red litmus paper, also available at drugstores. Dip a strip of red litmus paper into ammonia water, which is a base. Watch it turn blue. Try it on some other substances in the list you collected. Which give the alkaline test?

If you made some purple cabbage-water reagent, test it with a base. It will turn blue-green.

A GROUP TEST FOR SUSPICIOUS OBJECTS. There are thousands of group tests, far too many for even a fat reference book. Here's one more, for which you don't have the materials. But you can think up a situation around it.

This group test uses a reagent called *benzidine-nitroprusside.* With a drop of this reagent, a scientist can examine a clump of pale round objects under the microscope. Here they are.

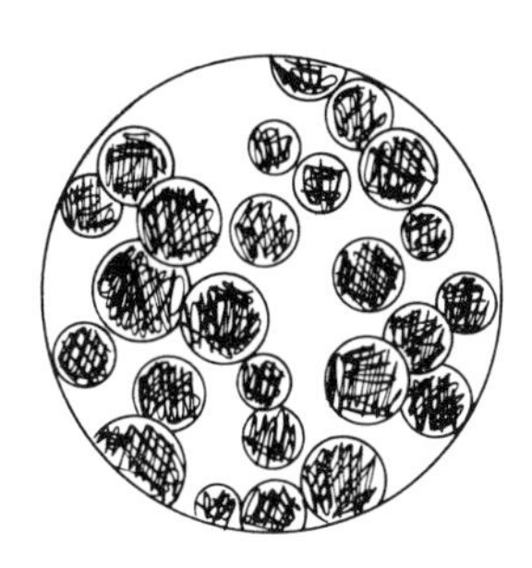
yeast

If the reagent produces no color change, the round objects are probably *yeast* cells. There are many kinds of yeast cells, found almost everywhere—in bread of course, and also on the skins of grapes and many other fruits, in dust, in soil.

If the reagent *does* produce a color change, to blue, the round objects are blood cells! There are many kinds of blood cells, from many kinds of living things.

FROM GROUPS TO SPECIFICS. Sometimes a group analysis is enough to get started upon. For instance, soil chemists and farmers know that potatoes grow best in slightly alkaline soil while beech trees do better in slightly acid soil. But for most purposes a group test doesn't give enough information. We don't just want to know that those little blue-stained circles are blood cells. We have another question: are they blood cells from a chicken—or a human being?

We need tests for *specific* substances.

SPECIFIC TESTS. The iodine test for starch is an example of a specific test. Put a pinch of starch (powdered or liquid) in a half glass of water and stir. Then add tincture of iodine, a drop at a time. The mixture will turn dark blue or purple.

Only starch turns color this way with iodine as the reagent. Check it yourself with sugar or salt. The substances are white like starch, but they are not starch. Then try a bit of bread, crackers, some egg white, a slice of potato. Which of them contain starch?

There are thousands of specific tests, far more than there are group tests. (Because a group contains more than one substance.) No scientist knows all the specific tests, or needs to know them all. That's what reference books are for.

Two Kinds of Analysis

So far, we have been asking "what" questions. *What's* in it? *What* is it? Sometimes the answer is a group answer: this liquid belongs to the acid group; these round things are blood cells. Sometimes it's a specific answer: this white powder is starch; these sparkly bits are gold.

This kind of analysis, that gives us a "what" answer, is called *qualitative analysis.* (*Qualis* is the Latin word meaning "of what kind.")

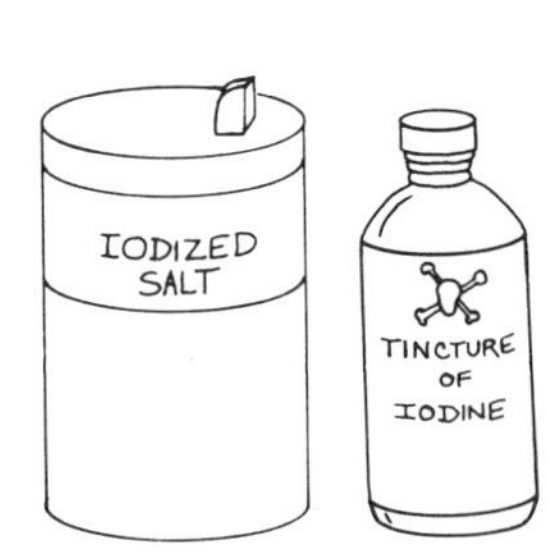

Sometimes a qualitative analysis gives enough information to a scientist examining an unknown substance. But sometimes it doesn't. Would you be willing to eat something that bears this label? CONTAINS IODINE. You need to know more, because this package and this bottle both contain iodine.

The tiny amount of iodine in a spoonful of iodized salt is healthful for your thyroid gland. (You also get iodine in ocean fish and other sea food.)

The large amount of iodine in a spoonful of tincture of iodine is enough to make you violently sick.

So, in many cases, we need to ask two questions: (1) *What* is in it? That's *qualitative* analysis. (2) *How much* of the "it" is in it? That's *quantitative* analysis. (*Quantus* is Latin for "how great.")

You've seen how the second question can be important in the dramatic matter of not swallowing iodine. It's important in everyday matters, too. For instance, here's the front of a cereal box, with splendid declarations in big type:

PEPPO! PEPPO! P-E·P-P·O!
Punchy Protein Pellets for Powerful People

That's a *qualitative* statement: there's protein in the cereal. But how much? On the side of the box there's a *quantitative* statement, required by law:

Protein	1 gram
Carbohydrates	24 grams
Fat	1 gram

Assuming that protein is desirable (it is) we want to know how Peppo compares with other cereals. Brand X cereal, for example, sells for the same price and displays these figures on the side of the box:

Protein	5 grams
Carbohydrates	20 grams
Fat	1 gram

These figures were obtained by quantitative analysis, using accurate weighing machines and precise reagents. Here's how you can do a *rough* quantitative analysis, to compare the amount of carbohydrates in two cereals. Starch is the main carbohydrate in cereals.

1. Weigh out equal small amounts of both cereals.
2. Crush each sample to a powder.
3. Put one sample into a half glass of water and stir.
4. Add iodine, drop by drop, until the mixture turns completely purple. How many drops were needed?
5. Now do steps 1 through 4 with the other cereal.

Suppose sample A took fourteen drops to turn purple while sample B took seven. Then sample A contains twice as much starch as sample B.

This drop-by-drop method of tagging the quantity of a substance is called *titration* (from the Latin for "label").

Iodine is a reagent. So is blue litmus, and red litmus, and *aqua regia* (the nitric acid and hydrochloric acid mixture used in the gold assay), and thousands of other substances used for finding out "what's in it?" and "how much is in it?" But reagents can't be used in every case. How could a reagent be used to determine what's in the sun? So scientists have invented all kinds of instruments for exploring matter in other ways. Let's look at a few.

Flame Tests

Why do the flickering flames of a campfire change color? What are the stars made of? Both questions can be answered by using the same method, called a *flame* test.

All substances are made of elements. Some examples of elements are: oxygen, carbon, and iron. When a substance is heated hot enough, it glows. It gives off light. Each kind of element within the substance gives off light of a certain color. The exact color of the light can help us identify the substance by telling us what elements are in it.

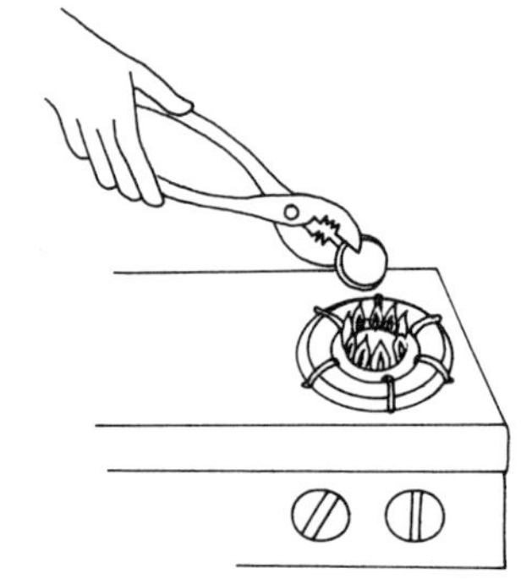

You may like to do a crude sort of flame test before you read about how scientists do the accurate ones. You will need a clean shiny dime, pliers, a gas or alcohol flame, a cup of water and a small metal pie plate. You will do the flame test on as many of these substances as you can obtain: table salt, boric acid, borax, alum, baking soda, washing soda, baking powder.

Hold the dime with pliers. Wet the dime with water, then touch it to the salt and hold it in the flame. You'll see a yellow glow. When the color is gone, wash the dime in water.

Try the flame test on borax or boric acid. The flame is green. Try the other substances in the same way.

Every single element gives off its own special color of light when heated. The element sodium gives off a yellow light. That's why the salt (sodium chloride) glows yellow. The element boron, in borax and boric acid, glows green.

There are more than a hundred elements, and each glows with its own special color! But you can't be sure of the difference between the apple green of barium and the slightly warmer green of boron. That's why the flame test done with a dime is very crude—just to give you a rough idea of what happens. Now let's see how scientists do it with a precision instrument called a *spectroscope.*

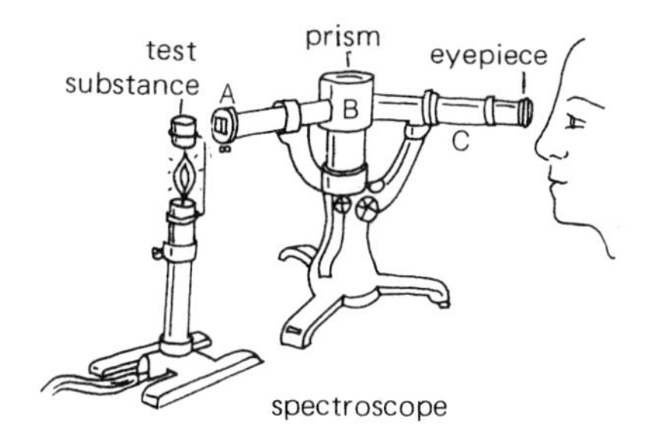

You can recognize the gas burner, shown here with a tiny cup to hold the substance being tested. The glow of the heated substance shines into tube A and passes through a glass prism (B). The prism spreads the light into its different colors. When we look through tube C we see this spread-out light.

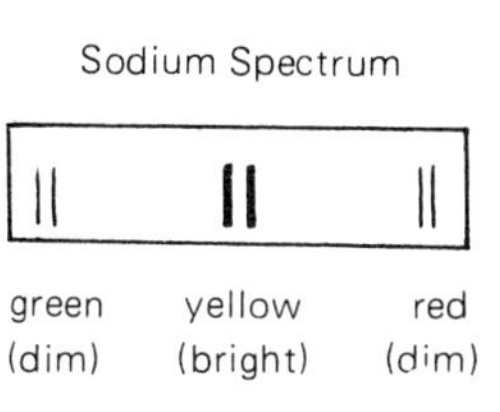

The spread-out light shows up as a set of colored lines. The light from sodium, for example, shows up as two bright yellow lines, close together, and four dim lines at the sides. From boron we get three lines, more spread apart, of different shades of green. Every element has its own private lines, its own "fingerprints."

You can make a simple spectroscope yourself.

Step 1. The very crude model. In a darkened room, shine a flashlight at a slant onto an LP record. Squint at the record, facing the flashlight. The fine grooves of the record break up the flashlight beam into a rainbow, as a prism does.

This very crude model demonstrates that the record can break up light into a spectrum. But you haven't tested an element yet. Now for step 2.

Step 2. The crude model. Cut a narrow slit in a card. Take the card and the record outside at night. Find a street lamp that shines with a blue-white light. This light comes from the element mercury in the lamp. Or find a street lamp that gives orange light, coming from the element argon. Or find a neon sign: neon gives a red light.

Hold the cardboard and record as in the illustration. Move your head a bit to one side or the other, until you see several colored lines. These are the "signature" lines of the glowing element in the lamp. No other element makes this exact group of lines!

Scientists don't use LP records for analyzing elements, of course. They use spectroscopes, as shown on the previous page, or similar instruments called *spectrometers.* The advantage of a spectrometer is that it has an illuminated scale to locate the exact position of each line.

Spectrometers have a wide range of uses. A

police scientist performs a flame test on a bit of food that may have caused a death by poison. He observes the flame through a spectrometer, looking for the lines that indicate arsenic. A chemist in a steel mill peers through a spectrometer at a glowing mass of molten iron ore in a furnace. The lines of colored light enable the chemist to identify the various elements in the ore. An astronomer attaches a spectrometer to a telescope and aims it at a star trillions of miles away. He photographs the hundreds of lines along the scale. They identify the elements present in the star.

There are almost no limits to the usefulness of a spectrometer. The substance being analyzed can be near or far, tiny or huge, a single element or many elements—as long as it can be made to give off light it can be analyzed and identified. Surely the spectrometer is one of the most valuable instruments of the scientist.

Colorimeters

You have probably seen pictures drawn a couple of centuries ago showing scientists staring thoughtfully into bottles of liquids. There isn't much you can tell by staring into a bottle of liquid, but maybe they were trying to estimate the *concentration* of the liquid. For instance, a liquid made of lots of blue copper sulfate crystals, dissolved in a little water, is said to be concentrated: its color is dark blue. If you dissolve just a few copper sulfate crystals in water, the liquid is less concentrated. The color is pale blue.

Such a way of testing concentration, by eye, is too rough for careful work. A more accurate way, in olden times, was to evaporate the water by heating gently. (Try it with saltwater.) The crystals were left behind and could be weighed. But this method took time and trouble.

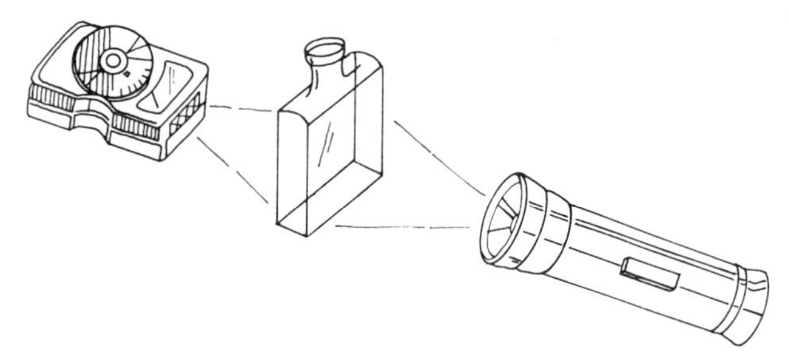

Nowadays there's an easier, quicker way. It came about with the invention of the photographic exposure meter. If you have one, you can try it yourself.

Shine a flashlight through an empty bottle. The light reaches the meter, which measures the brightness.

Now, if you fill the bottle with a concentrated solution (dark coffee, for instance) much of the light will be held back; the meter will show a lower reading. If you use a less concentrated (more diluted) solution (such as weak coffee) the meter will show a higher reading.

You may wish to prepare a set of test bottles, with instant coffee crystals or with food coloring. The first bottle of the set should contain clear water. The last bottle should contain the highest concentration you can make. The other bottles should contain in-between concentrations. Keep a record of the amount of dissolved material in each bottle, and the reading of the exposure meter with each. Then let somebody prepare an "unknown" sample, without telling you the concentration. Try to determine it with your meter and test bottles.

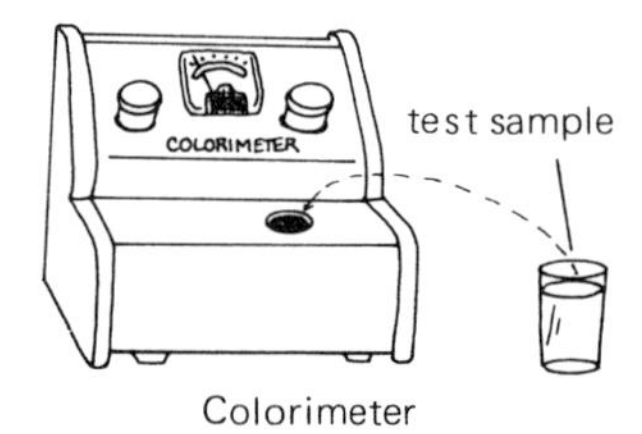

Colorimeter

The professional instrument of this type is called a *photoelectric colorimeter.*

"Photo" comes from the Greek word for light. "Electric" refers to the meter's working by electricity. The "color" part of colorimeter refers to *two* factors: (1) The deepness of the color is a measure of the concentration. That's no surprise—you found that out in the previous experiment. (2) (This *is* a surprise!) If two or more differently colored substances are dissolved in the same water, it is possible to measure the separate concentration of each!

For instance, suppose you want to test a liquid mixture containing blue copper sulfate *and* orange potassium dichromate. Just put a blue filter, such as blue cellophane, in front of the bottle. This allows blue light, passed by the copper sulfate, to register on the meter. But the blue filter keeps back other colors. Then change to an orange filter and get a reading for potassium dichromate. Lots easier than staring into a bottle—and lots more accurate.

Crystallography

Here's a cardboard mystery box that somebody hands you. There's a big prize if you guess what's inside. The only clue: it's quite heavy. The only restriction: you can't open the box. How to go about it? Here's an imaginary way, in three steps.

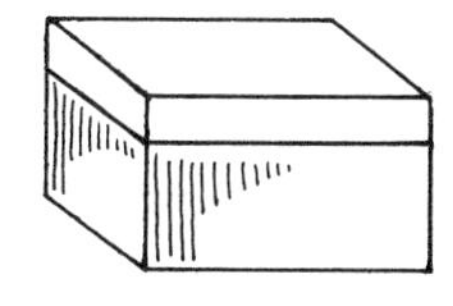

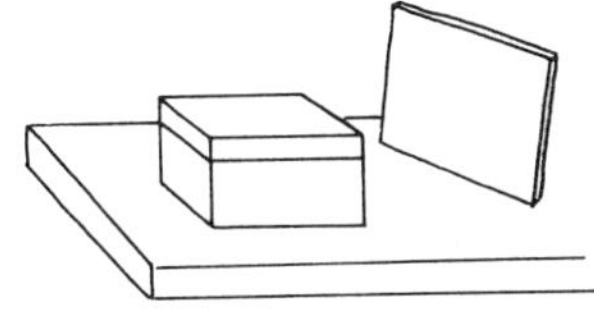

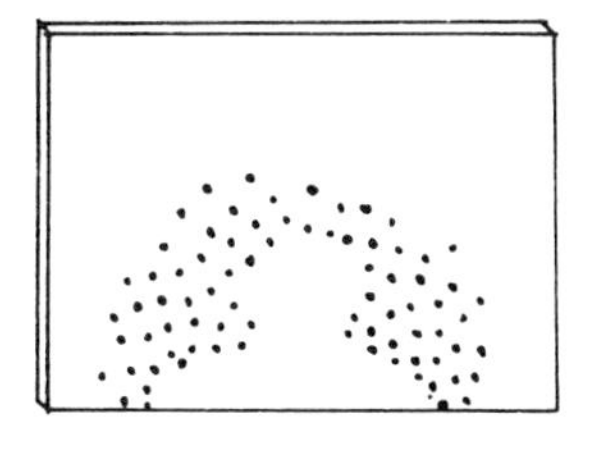

1. Put a sheet of cardboard behind the box. Shoot a shotgun at it from several feet in front. (A shotgun shell contains lots of little balls of lead shot.) The balls of lead shot are stopped by the mystery object, but not by empty spaces in the box. The result is a cardboard peppered with holes, in a pattern like this.
2. Turn the box to one side, put fresh card-

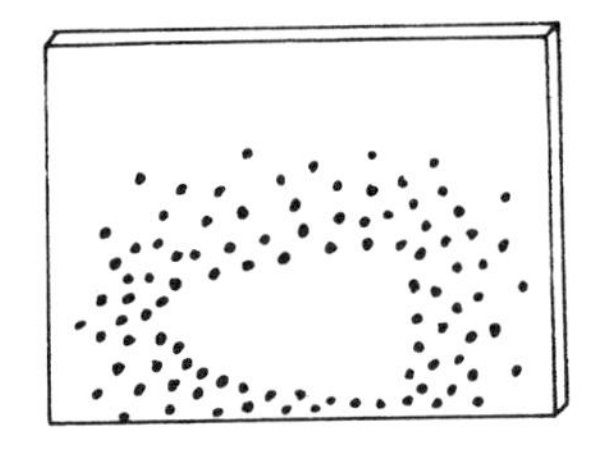

board behind it, and shoot another shell. Result: another cardboard peppered with holes.

3. Turn the box again. Another cardboard, another shot, another pattern of holes.

By now you've guessed what's inside. But what do electric irons and shotguns have to do with scientific instruments?

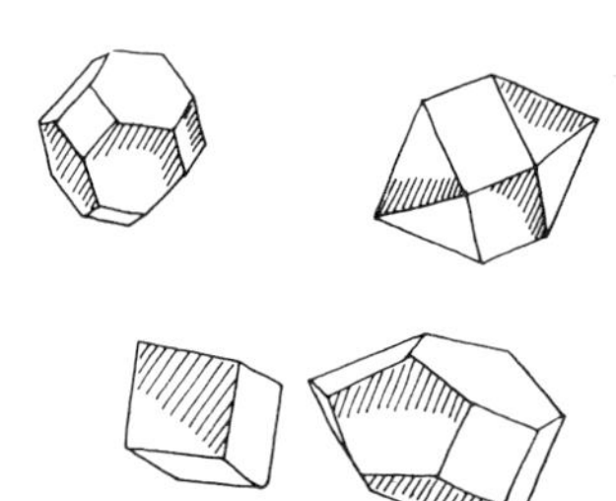

The shotgun method is used for finding out about the insides of *crystals.* Most solids are crystals. You can see the crystalline shapes of salt, sugar, sand, diamonds, because the crystals are big enough to see with the naked eye or a magnifier. But the penny in your pocket is also composed of crystals—copper crystals too tiny to be seen without a microscope. Crystals make up the black lead in your pencil and the white enamel of your teeth.

To really understand a substance, a scientist has to understand the structure of its crystals. Why does copper conduct electricity, while copper-colored plastic doesn't? Why do living teeth get cavities, while artificial teeth don't?

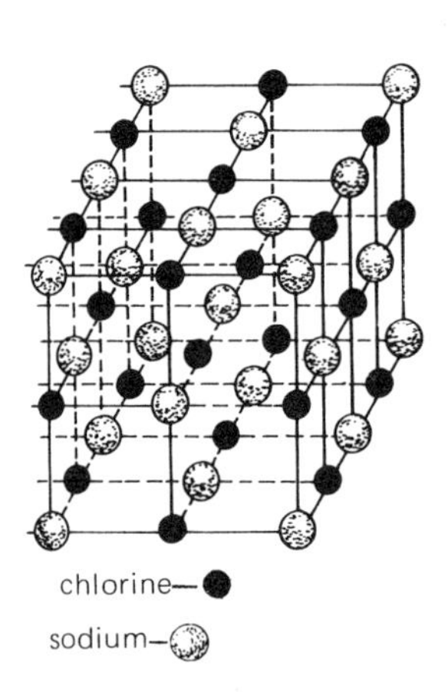

The study of crystals is called *crystallography.* The instrument used is an X-ray diffractometer. Let's see what it looks like, how it works, and what it has to do with shotguns.

This is a diagram of a tiny crystal of table salt (sodium chloride). Such a crystal is much too small to be seen even with the most powerful microscope. Each grain of salt in your salt shaker is made of billions of crystals, neatly stacked together.

How do we know? If we can't see them, how do we know they are arranged like this? Because we have the "shotgun" method to show us what we can't see.

The "shotgun" is a device that sends out a stream of tiny X-ray "bullets." Those that reach empty spaces in the crystal zip right through. They strike the "cardboard" which is a photographic film. Those "bullets" that strike atomic particles (the dots in the salt diagram) bounce (diffract) in a new direction. The direction depends on the angle at which they strike. Some bounce backwards or sideways. Some bounce forward at a slant and strike the film. When the film is developed, we see a pattern of spots, made by the X-ray bullets.

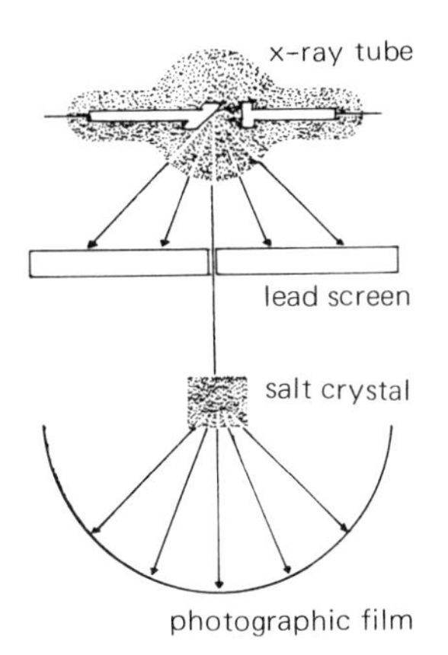

Next, we turn the crystal to a new position and shoot again, on a new film. Then another turn and another shot, and still another at a new position, and still another. It isn't nearly as easy as shooting at a box and guessing what's in it, because each atom is really a separate little target. In fact, the patterns are much too complicated to figure out just by shape. It takes a computer to do the job.

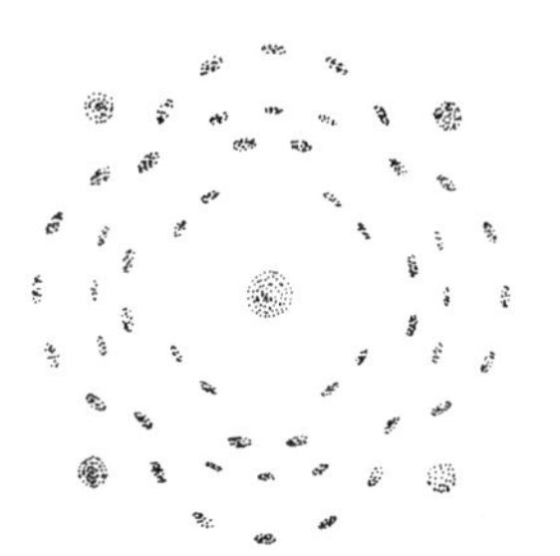

Look at the diagram of a salt crystal on the previous page, and its X-ray on this page. Salt is one of the simplest crystals. Now look at one of the most complicated.

This is a diagram of a single molecule of DNA, whose full name is almost as complicated as the diagram—*deoxyribonucleic acid.* Can you imagine all the X-ray photography, all the calculations by computer, all the thinking that went into figuring out the structure of this molecule? This work was done by four scientists: Rosalind Franklin, James Watson, Francis Crick and Maurice Wilkins.

DNA molecules are present in all living things, in almost every single cell. They make up the structures called genes, which do two kinds of jobs.

D N A Molecule

1. Genes carry heredity from parents to offspring. You resemble your parents because their genes for eye color, nose shape, limb length and millions of other traits were passed on to you.

2. Genes direct the behavior of your cells. The genes in the bone cells of a growing child cause more bone material to be formed. The genes in muscle cells make the muscle able to contract and expand. The genes in nerve cells make nerves able to transmit messages.

Cultures

Had a cold lately? It was caused by a tiny living thing, a microorganism called a *virus.*

Had a piece of cheese lately? It was made out of milk by microorganisms called *bacteria.*

Had a slice of bread lately? It was made from flour by microorganisms called yeast cells.

Had a case of malaria lately? (Not very likely, thanks to science!) Malaria is caused by microorganisms called *protozoa.*

There are thousands of different kinds of microorganisms. A few are harmful to other living things; most are harmless or useful. All are extremely interesting objects of study to scientists.

Studying microorganisms is no easy matter. To begin with, they are small and must be looked for with microscopes. And then, when you do find them, it's hard to recognize them.

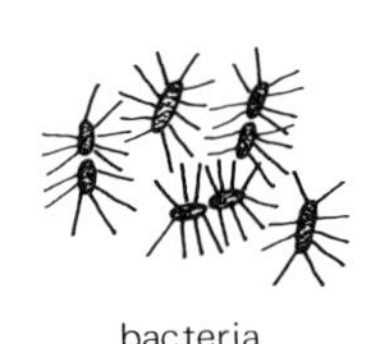

bacteria

These little rod-shaped things might be the harmless bacteria that live in people's mouths—*or* they might be the germs of tuberculosis.

Some little wiggly objects might be *E. coli,* the harmless bacteria that live in people's intestines—*or,* they might be the germs of typhoid fever.

You can't easily catch microorganisms on the fly, as they drift through the air, or as they drift through your blood. You can look hard, but you won't discern the several thousand bacteria on each of your fingertips, or the several million yeast cells on the skin of a single grape. But scientists have developed a system called *culturing*, for making microorganisms easy to find and study.

Culturing

To culture a microorganism, you provide it with the right conditions for a healthy life. Good health comes from having the right food and drink, the right place to live, the right temperature. Each kind of microorganism has its own definition of right, of course.

Under the proper conditions, a microorganism will grow, divide in two, grow, divide again and become four, and so on—dividing and growing and doubling, over and over again. In a very short time a single microorganism, almost impossible to locate, has become a busy culture of several trillion, handy for close study.

Maybe your doctor has done a throat culture for you. He swabs your throat with a bit of cotton at the end of a stick. Then he stabs the cotton into a tube or dish containing a jellylike substance, a *culture medium.* This is the "right food and drink" and the "right place to live" for the kind of microorganism the doctor is looking for.

Now for the right temperature. The doctor places the tube or dish in a heated cabinet, an incubator, which is kept at body temperature. (That's the proper temperature for microorganisms that live in your body.) In a few hours the single cells have multiplied into whole colonies. Here are a few such colonies.

Some microorganisms are readily recognized by size and shape. But most have to be identified by further tests. The first is usually a group test, done with a reagent called *Gram's stain.* This group test causes some microorganisms to turn violet. These are called *Gram-positive.* Others turn red, and are called *Gram-negative.* (Gram-positives are usually killed by penicillin, and Gram-negatives by streptomycin.)

The Gram's stain test is a bit complicated, and you probably can't get the reagent at your drugstore. But it's easy to grow cultures at home. All you need is a slice of boiled potato for the culture medium.

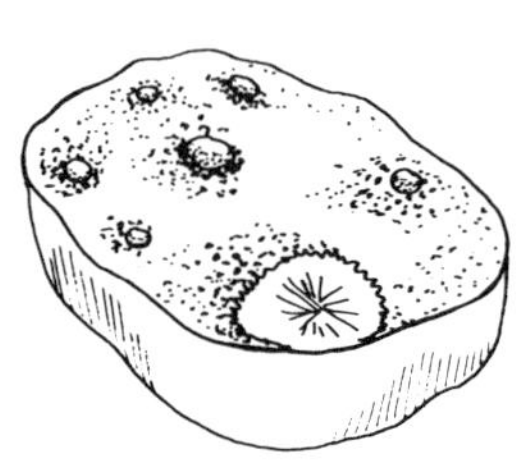
potato with several colonies

Put the potato in a small bowl. Sprinkle it with a pinch of dust. Cover it to keep the potato moist, and wait for results. In a few days you may get several of these colonies.

Tracers

Breakfast: grapefruit, scrambled eggs, buttered toast. Down goes the breakfast into your digestive system, and off you go on your day's business.

And then what? Exactly where does each substance in the food go? How long does it take to get there? How is it used? Are the answers different for a sick person than for a well person?

And so on and so on. Tough questions to answer—and for a long time the answers were mainly guesswork. It isn't easy to trace substances through a living thing without disturbing it to a point where the answers are not reliable.

That was the situation until the discovery of *radioactive tracers.* These are substances that give off invisible rays. Scientists can detect and count these rays with an instrument called a Geiger-Müller counter. Let's see how such an instrument, together with radioactive tracers, can be used to answer the question: "What happened to breakfast?"

Iodine is an important substance for good health. It's used by the thyroid gland, which is located on the front and sides of the Adam's apple. Some foods, especially seafoods, contain iodine. Questions: (1) Does the iodine in food get to the thyroid gland? (2) If so, how long does it take? (3) Does the iodine also get to the rest of the body?

To find the answers, scientists prepared special iodized salt. This salt contains iodine that has been made *radioactive*. (For more about radioactivity, see page 57.) The salt was put into food and eaten. Then the Geiger-Müller counter was held near the thyroid gland and other parts of the body. The counter made clicking sounds, sending out its signals wherever radioactive iodine was detected. Answers: (1) The iodine was detected in the thyroid gland. (2) The trip from stomach into blood into thyroid gland took several minutes. (3) The iodine also circulated through rest of the body but was not absorbed.

Geiger-Muller Counter

Iodine is only one of the many substances that can be made radioactive. Sodium, calcium, carbon, nitrogen, hydrogen, iron, and many others are used as tracers for finding out how living things work.

You can probably think up a few more uses for radioactive tracers. For example, when we read advertisements about iron pills that give you a lot more energy, we may want to find out if the pills will work. A qualitative analysis shows that the pills do indeed contain iron, and a quantitative test shows lots of iron per pill. There's still a question: how do we know that the iron is absorbed and used? To find the answer, the scientist prepares a food or drink containing radioactive iron, and traces its path through the body with a Geiger-Müller counter.

Tracers are valuable for other purposes, too. Suppose there's a break somewhere in a sewer pipe. Disease germs are seeping into the soil and finding their way into the town water supply. To fix the break we have to dig up the streets—but where? Every excavation means blocked traffic and money spent for chopping up and repairing the pavement.

Tracers to the rescue! Pour some radioactive chemicals into a toilet. Flush the toilet and trace the radioactivity with a Geiger-Müller counter. Wherever it follows a map of the sewer system, fine. But where it shows up in the soil outside of the pipe—there's the break.

THE STRENGTH OF MATERIALS

How Strong Is It?

That's a question about matter, and everybody asks it at some time or other. You might inquire about a pair of sneakers, or a tennis racket, or the cables of a fully loaded elevator you're riding in. A dentist might pose it about a gold alloy metal for making tooth inlays. An engineer might ask it about a steel girder for a bridge.

The strength of materials is such an important subject that there's a whole branch of science and engineering dealing with it.

What do we mean by strength? Here's a brick-and-cement bridge, built by the Romans, still carrying traffic—a heavier load of traffic than when it was built two thousand

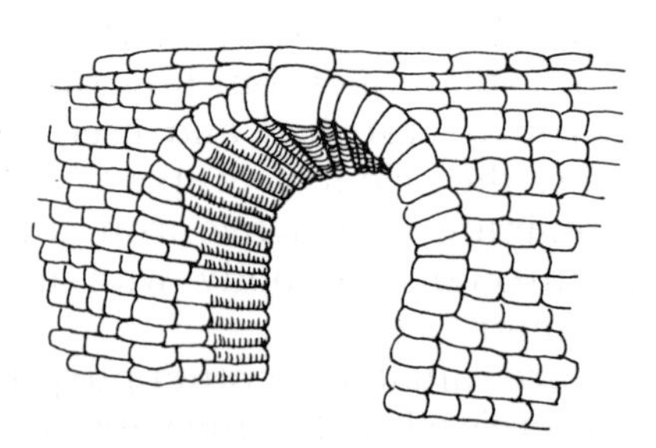

Roman Arch Bridge

years ago. The load rolls along on top of the brick and cement, and *presses* down on it. Brick and cement are very strong under com*pression*.

The roadway of the Brooklyn Bridge hangs from steel cables. The load *tenses* the cables. Steel is strong under *tension*. The cables hang from stone piers and press down on them. Stone is strong under compression.

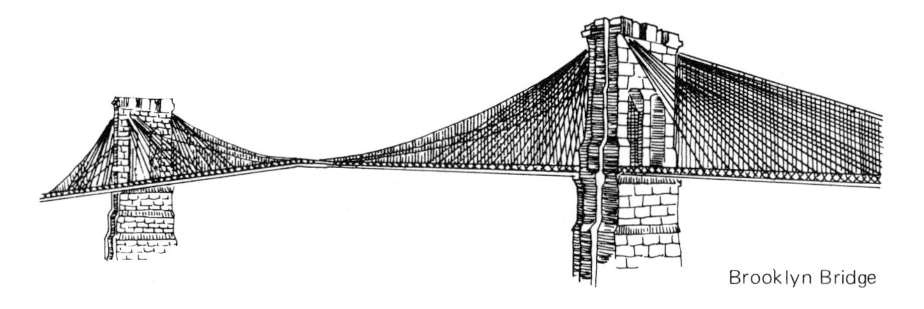
Brooklyn Bridge

Clearly, we have to match the materials to the job they do.

The job of a guitar string—to make musical sounds while under tension—calls for the use of steel (or nylon, or copper). But the job of a xylophone bar—to make musical sounds while being struck (compressed) by a hammer—is better served by using wood.

When you buy a roll of picture wire or a box of tacks, you don't have to test the wire for tension, or the tacks for compression. Somebody else, a mechanical engineer or a metallurgical engineer, has already done that for you.

Testing Strength

Testing is done with machines that can measure a material for tension, compression or for still other kinds of strength. The machine shown here is a compression tester, being used on a brick. The handle on top is turned,

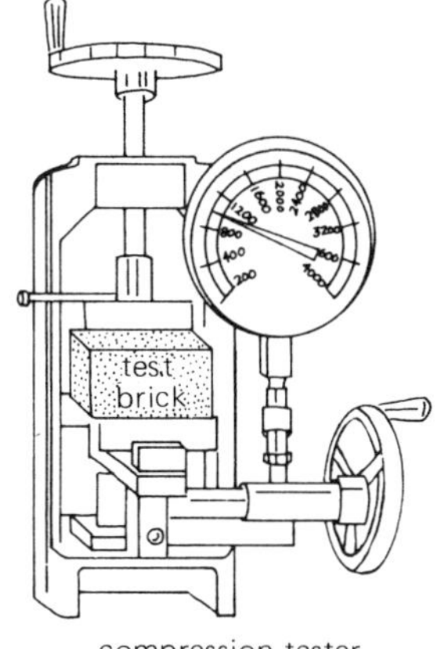

compression tester

forcing a steel bar down against the brick, until it cracks in two. The force needed to accomplish this is registered on the dial. The lower wheel, at the right, is used in testing for shear, which you will read about later in this chapter.

This testing machine is used on materials of great strength, such as iron, steel, and concrete. Other testing machines are used for weaker materials. Paper manufacturers, for example, use such machines to test their products under *two* conditions of tension—dry and wet. You can see how that matters to a user of paper bags and paper towels.

You can test some paper samples yourself. You will use a crude testing machine, but the results would be almost acceptable to a paper manufacturer.

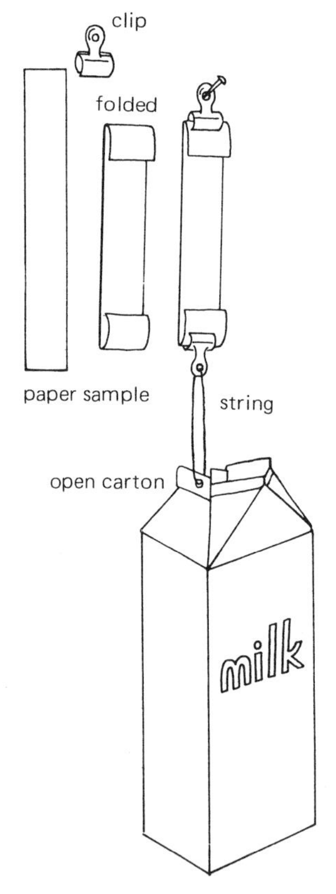

Cut strips of various kinds of paper (newspaper, typing, napkin, toweling, wax paper, etc.) one-half inch wide and six inches long. Fold over each end. Fasten a little spring clip to each end of one strip. Hang it up as in the diagram. Hang an empty milk carton to the lower end. Place a basin underneath, and pour water slowly into the carton until the paper breaks. You can determine how much of a load caused the break by watching the water level in the carton. A full quart carton of water weighs about two pounds.

Now repeat the test with another sample of the same kind of paper, but first dip the middle two inches in water. This test will give the wet strength of the paper. Would you guess that the wet strength of tissue paper is much lower than the dry strength? How about wax paper? Do you think the length of the strip makes a difference? Suppose you twisted the strip into a tight coil, would that make a difference? Could that explain why ropes are made of twisted cords?

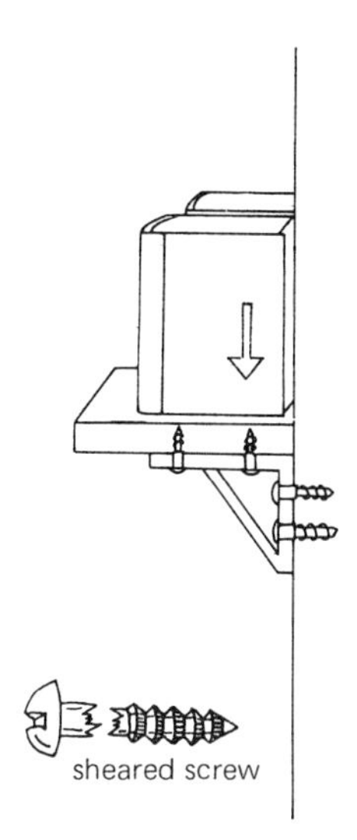

SHEAR. Another test for the strength of materials is the *shear* test. The word "shear" reminds you of scissors. Think of the action of cutting a sheet of paper with scissors. The blades slide past each other. They force one part of the paper to slide sideways, to shear away, from the other.

Shearing doesn't happen only to paper or cloth. For instance, look at these steel screws holding a bookshelf to the wall. If the screws were weak in shear strength, they would shear apart. The head end would separate from the tail end, like this.

The same machine that tests materials for compression and tension is used for testing shear strength.

FATIGUE. Sometimes a material seems to just get tired. This tiredness is actually called *fatigue*, and it happens mostly to metals. You can see it happen to an inch-wide strip of metal, about three inches long. It's easy to cut (or rather, shear) such a strip from an aluminum beverage can, using tin snips or an old pair of scissors.

Grab each end with a pair of pliers. Try to pull the strip apart. Any luck?

Then bend the strip back and forth, doubling it on itself, several times, to tire it out. Now it's easy to pull it apart!

Metal fatigue worries people who work with metals—automobile engineers, airplane designers, bridge constructors—because it isn't well understood or easily tested. It's not caused by an over-*load*, like too much weight hanging from an elevator cable, but by over-*use*. And how much use is overuse? It varies from Sample 1 to Sample 2 to Sample 3 of the same batch of metal.

The science of metallurgy, like every other science, is far from complete. There's much to be learned about metals. That's why commercial airplanes are periodically tested for fatigue. An inspector goes over the airplane part by part, rivet by rivet, looking for the tiny hairline cracks that signal the mysterious tiredness called metal fatigue.

Corrosion Resistance

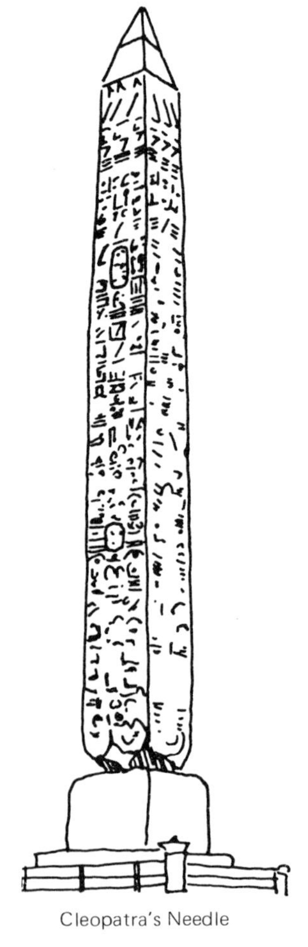
Cleopatra's Needle

In 1897 an Egyptian obelisk was brought to Central Park in New York City. The obelisk had been standing in the clean dry air of the Egyptian desert for over two thousand years, yet the carved figures and inscriptions were still clear and sharp. Today, after less than a century in New York air, the carving is almost all gone. It has been corroded by the chemicals in smoke and auto exhaust fumes.

Resistance to corrosion matters very much to a sculptor, choosing a material for an outdoor statue. It matters to architects designing buildings, to engineers designing cars, to ladies buying stockings. As air pollution increases, so does the cost of corrosion.

Cutting down on air pollution is an obvious and necessary step, if we're to protect our homes, our clothes, our lungs. But we also need to choose and develop materials that resist the effects of pollution. That's why scientists—especially chemists—are testing materials and creating new ones.

If you have access to polluted air (and who hasn't, these days?) you might want to try a few simple corrosion tests. Get two of each of the following: steel nails or screws, brass screws or cup hooks, plastic spoons, stainless steel spoons, three-inch squares of used nylon stocking, squares of aluminum foil, shiny pennies.

Put one of each pair of items in a plastic bag and seal the end with scotch tape. Put the other items in another plastic bag but leave the end open to the air. Hang the two bags on a line or pole, or fasten them firmly to a windowsill. Wait a few weeks for the corrosion caused by polluted air. Which items became corroded? Which items did not?

Air pollution is not the only thing that can cause corrosion. Boat designers worry about seawater and salt spray. Designers of telephone cables have to pay attention to the chemicals in damp soil. Furniture manufacturers test for the effects of vinegar and alcohol on kitchen tables. Corrosion is all over the place!

There are many more tests, for different properties of materials. Some are easy to do and interesting. You may wish to try the following.

Surface Hardness

When a rubber ball is dropped on a hard surface, it bounces higher than when it's dropped on a soft surface. This is the basis of one kind of surface hardness test. Some parts of machines, certain kinds of glass, some kinds of building materials need to have hard surfaces. The test is done, not with a rubber ball but with a tiny diamond-tipped weight. Diamond is the hardest material of all.

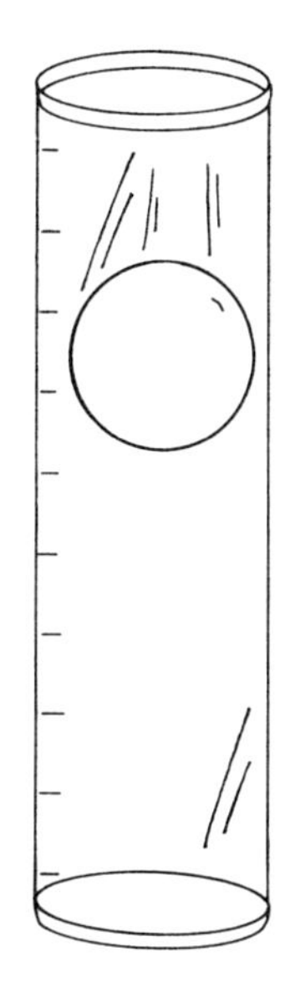

You can do a rough kind of surface hardness test with a Ping-Pong ball and a transparent plastic tube used for packaging household items. Or you can roll a tube out of sheet plastic.

Cut the tube to a length of ten inches. With a felt tip pen, mark inches on the side. Place the tube over the surface to be tested—a porcelain dish, for example. Hold the Ping-Pong ball at the top and let it fall. How high does it bounce back? Let's say seven inches. That's seven-tenths as far as it fell. The surface hardness of the porcelain is .7.

Try the test on other surfaces—glass, various kinds of plastics and woods. The test works fairly well on hard surfaces (it would work much better with a diamond ball but—). On cloth and other soft surfaces you'll get nothing but a dull thud.

Brittleness Test

You have had many proofs that glass is brittle, and you have often swept up the proofs. Brittleness doesn't matter too much in a drinking glass, but it's disastrous in an auto windshield. So, in glass factories, samples of glass are tested for brittleness with an *impact tester.*

Let's see how it works.

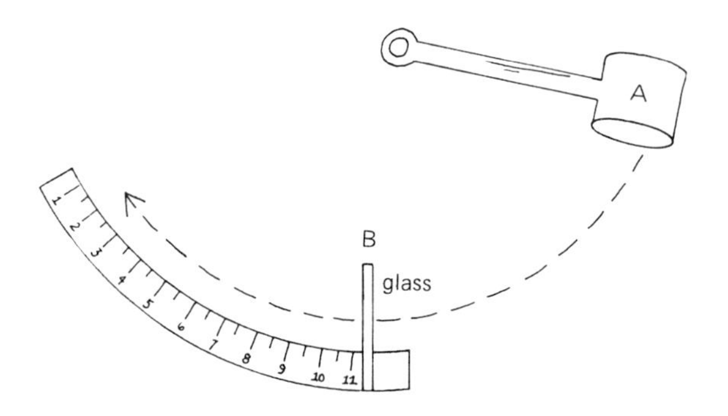

A steel hammer is dropped from position (A). It swings down and strikes the glass sample (B). The glass is shattered and the hammer keeps going on and upward in an arc. The high point of that arc tells us about the brittleness of the glass.

The more brittle the glass, the less force is needed to shatter it. Less energy is taken out of the falling hammer. Therefore, more energy is left to keep the hammer going after it smashes the glass, and the greater the arc becomes. But with tough glass, more energy is needed to smash it, less energy is left to keep the hammer going, and the smaller the arc becomes.

That's how they measure the brittleness of glass. The glass for a store window needs to be very strong. For an ordinary window it can be less strong. The glass for a framed picture can be quite brittle.

ANNEALING. Why not make all glass as strong as possible? Because brittle glass is cheaper to make. When molten glass is poured into a shape and allowed to cool quickly, it's brittle. If instead it's allowed to cool very slowly, for hours or days, it's stronger, but the process costs more. This process is called *annealing.* The glass for the Mount Palomar telescope was annealed in a slowly cooling oven for ten months!

Other materials can be annealed in the same way. Try it with two steel sewing needles. With pliers, hold one needle in a candle flame until it's red hot. Then quickly cool it in cold water. Heat the other needle red hot, then very slowly draw it out of the flame.

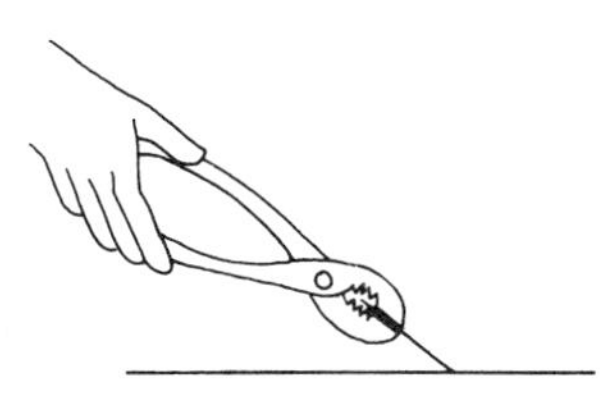

Test each needle this way: hold one end with pliers and press it at a slant against a hard surface such as a dish. The quickly cooled needle is brittle, and will probably snap in two. What happens to the annealed needle?

Metalworkers and glassworkers have known about annealing for many centuries, but they didn't know why it happened. It was a trial and error process for each new material.

The same is true of other processes that increase tensile strength, or surface hardness, or corrosion resistance, and other desirable properties. But the knowledge is coming along, through the work of scientists—metallurgists, chemists, and especially crystallographers.

Let's see why the work of crystallographers is especially important.

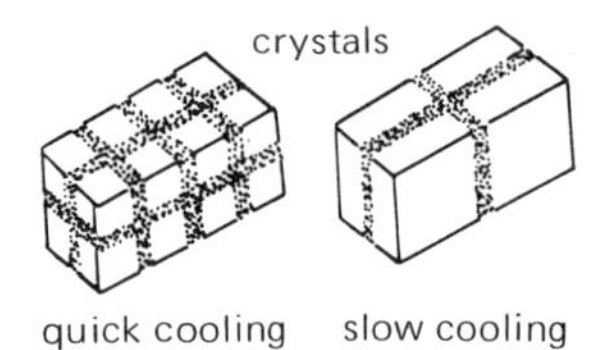

Most hard solids (such as metals and rocks) are formed of crystals. The crystals, in turn, are made of atoms. The forces that hold together the atoms *inside* a crystal are very strong. But the forces *between* crystals—that bind a crystal to its neighboring crystals—are not nearly as strong. When a heated metal is annealed by cooling slowly, it forms large crystals. There are few in-between weak spaces. When a heated metal is cooled quickly, it forms small crystals with lots of in-between weak spaces.

The closer we look, the more we see!

2

time

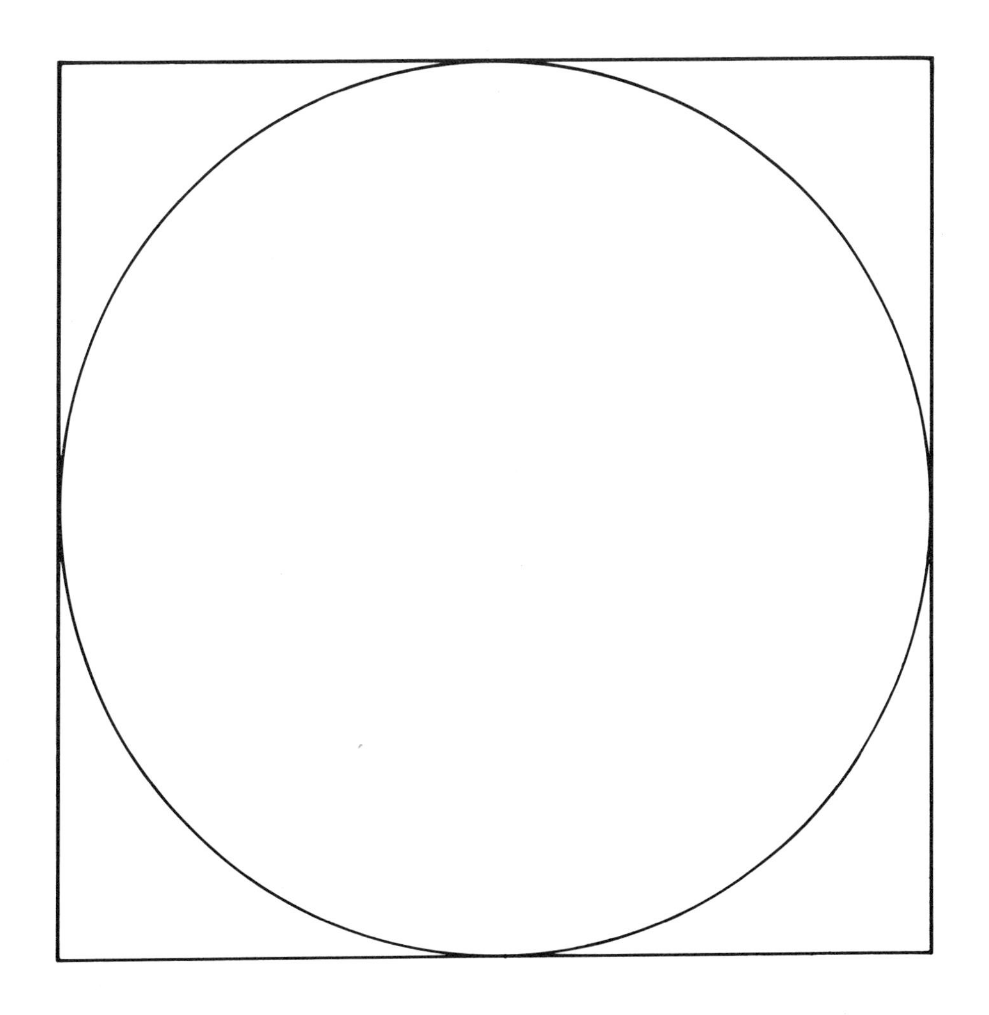

What's the smallest amount of time that matters in your daily life? Maybe a minute or so—the time it takes to be late for the bus, or on time for a TV program, or just right for the difference between a soft-boiled egg and a medium-boiled egg. An ordinary clock or watch takes care of your time problems quite satisfactorily.

How about time for a caveman? Maybe an hour or so—the time to turn back after a day of hunting and foodgathering, to return to the cave before nightfall. Observing the position of the sun in the sky gave him a rough idea of the time of day. Adding day after day gave him a record of the time of year.

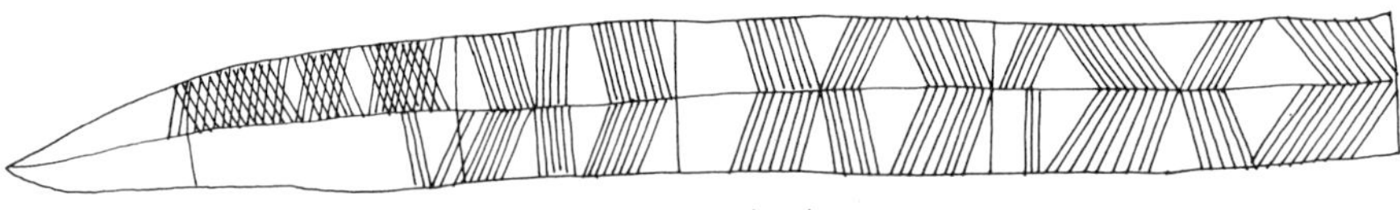

caveman calendar

To a crew of astronauts, returning from space, time is a split-second matter. Their capsule hits the top of the earth's atmosphere at a speed of five miles *per second.* From there on, if everything has been timed right, they continue to a safe landing. The timing is much too critical for a human being with a stopwatch. A quartz-crystal clock, accurate to a thousandth of a second per month, keeps time for the space capsule's machinery.

How about a hundred years? That seems like an easy block of time to measure. And it is easy, if you're counting present-day time. You just use up a hundred calendars, one after the other. But it's not easy for a scientist examining the ashes in an ancient cave, asking "When did that fire go out?" To answer that question with an accuracy of a hundred years more or less, the scientist has to use a complicated series of tests called radiocarbon dating.

So let's take some time investigating time.

What Is Time?

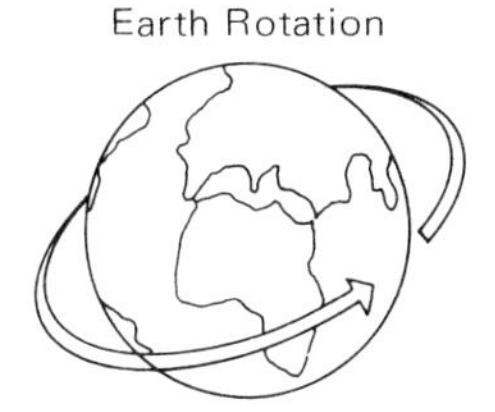

Of course you know what time is. It's what you count with a clock or calendar. It's what the dictionary describes as "the interval between two events." Today's sunrise is an event. Tomorrow's sunrise will be an event. The approximate interval between the two events is called a *day*. You can count on it because the cause of sunrise, the earth's rotation on its axis, is steady and reliable. It is always the same interval of time.

In the same way, we have a time interval called a year. This interval is determined by the position of the earth in its orbit around the sun. Today the earth is in a certain place in its orbit. That's Event number 1. The next time the earth is in that certain place will be Event number 2. The interval between the two events will be a unit of time called a *year*.

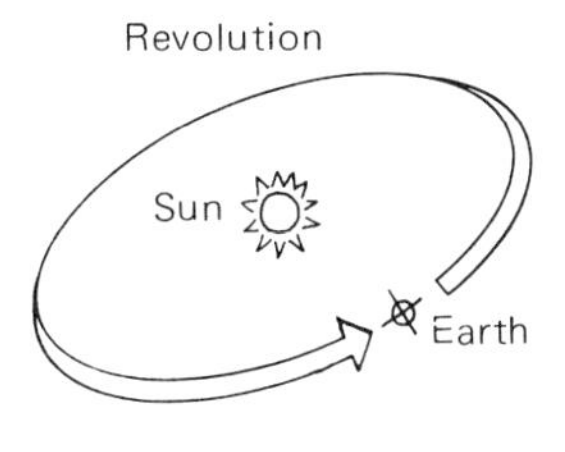

Here are two more events: (1) you get into a bus; (2) you get out of the bus. Although these are two events, the interval between them is *not* a unit of time because it's not always the same. The bus doesn't always go at the same speed; you don't always get in and out at the same two places; and so on. If you want to know how much time you spent on the bus, you have to separate the bus ride interval into smaller units that are all alike, that can be counted, and that everybody agrees on.

Everybody agrees on a time unit called a *second*. Everybody agrees that sixty seconds make another time unit called a *minute*. Everybody agrees that sixty minutes make another time unit called an *hour*, and twenty-four hours make a day.

Time Counters

Thus far, everything is agreeable, so we can carry a time counter called a watch. We can look at the watch during Event 1 (getting on

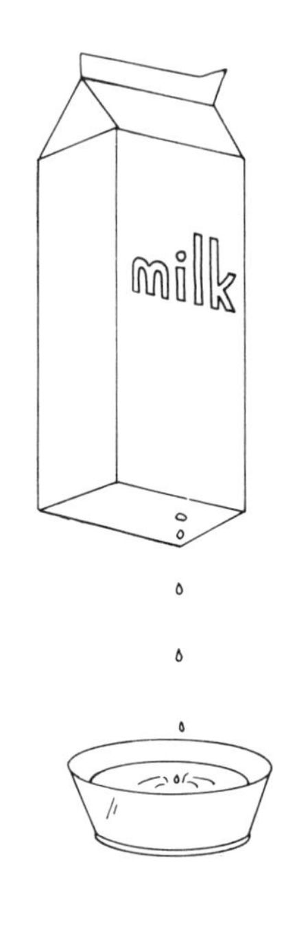

the bus) and again during Event 2 (getting off the bus) and announce that the bus trip took two hours, seventeen minutes and twenty-three seconds.

But there's a hitch to all this. We first have to agree on an interval called a second. We have to find or make two events that are one second apart. These two events have to be the same anywhere in the world. We have to be able to repeat them, second after second, time after time.

Put yourself back in time, to the time of the Greeks a couple of thousand years ago, and face the problem. Find something, or make something, that produces *equal* time intervals, again and again.

WATER CLOCKS. How about the dripping from a tiny hole in the bottom of a jar of water? You'll probably agree on that one and then think again about such a second. Won't the drip be faster when the jar is full than when the water level goes down? Test it with a milk carton. Stick a pinhole near the bottom. Fill the carton with water. Time the drip of ten drops with the carton full, half full and almost empty. What happens?

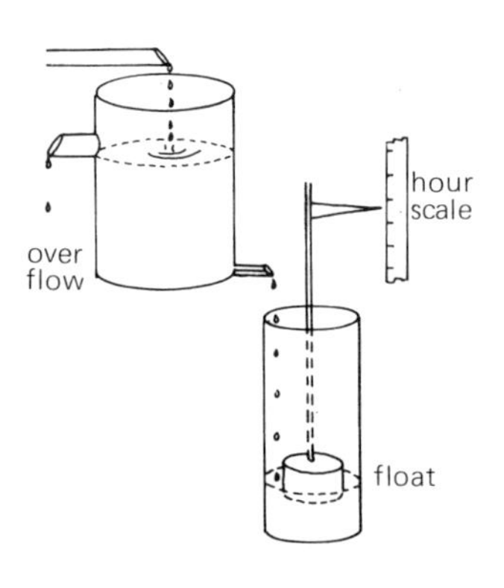

The Greeks thought about that too, so they rigged up an arrangement to keep the jar full of water at all times. Here it is: a float, a rod, a pointer and a scale. It's a water clock called a *clepsydra*, which in Greek means "a thief of water."

The clepsydra worked well enough for timing events that took a fairly long time, such as the 26¼ mile race called the marathon. But even for such events it wasn't altogether reliable. For instance, a change in temperature might change the time interval between drops. You might want to test that idea with a milk carton, using cold water and warm water. You could use a regular clock or watch for comparison (which the Greeks couldn't).

SAND GLASS. Here's an improvement. A sand glass keeps time by the fall of sand from the upper glass, through a narrow neck, into the lower glass. An hourglass is a sand glass with enough sand to take a full hour for the complete job. One problem: no two hourglasses keep exactly the same time intervals. That's just about impossible, because they would need to contain the same number of sand grains, and the necks of the glasses would have to be exactly the same size.

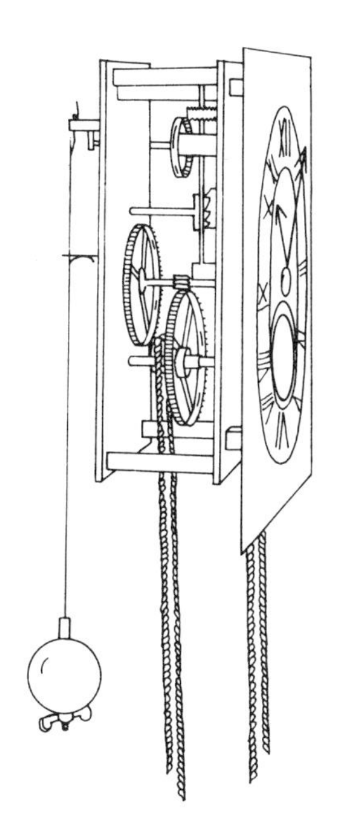

PENDULUM. Back to the search for equal time intervals. It is related that in 1583 a scientist named Galileo stood in the rear of the cathedral of Pisa, in Italy, during church services. He watched a huge chandelier that hung from the ceiling (and which still hangs there). A breeze had set it swinging slowly and gently. The swings *seemed* to take equal intervals of time. To check it, Galileo counted the pulse beats in his wrist during each swing. The swings did take equal time intervals or equal numbers of pulse beats per swing. Gradually the swings became smaller and smaller, but the time interval of each swing seemed to remain the same! (Try it with a book and a string.)

Getting somewhere at last! A hanging object seems to swing within equal time intervals. Such an object is called a *pendulum,* which comes from the Latin meaning "hanging thing." This pendulum clock, made in 1656, has a pendulum, of course. A set of weights keeps the pendulum swinging. A set of gears and hands count the swings and change them into hours, minutes and seconds.

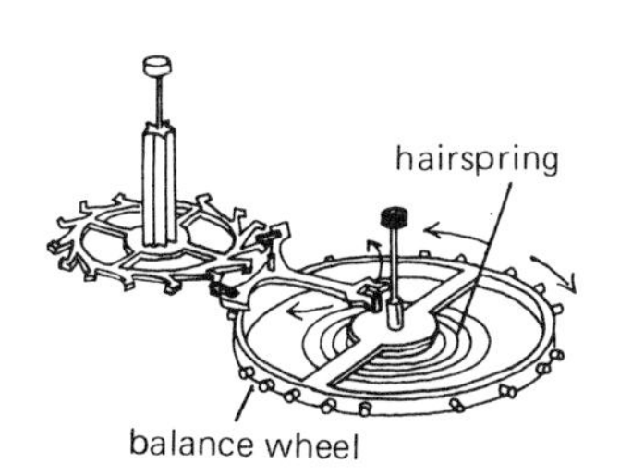

HAIRSPRING. There's no pendulum in a wristwatch, of course. Most watches and clocks have another device for dividing time into equal intervals. It's called a hairspring. It coils and uncoils over and over again, driven by a spring that you wind up daily. The coil-uncoil of the hairspring causes a little wheel to turn back and forth. The wheel turns a set of gears and hands that count off the seconds, minutes, and hours. This is much handier than a pendulum clock for most people, especially for airplane pilots and scuba divers.

Hairspring watches are the most commonly used timekeepers. They're small, easy to carry on the wrist or in the pocket, and even the cheapest ones are fairly accurate. The very good ones, which are also very expensive, are accurate to within a few seconds per day.

But even these clocks have problems. For instance, placed flat on their backs, they run a tiny bit faster than when they are set on edge. You could solve that problem by leaving the watch undisturbed on a shelf, or by holding your arm in one position, night and day. Most people prefer to wear or to carry them, and to move their arms when they please. So, if we are searching for a still more accurate timepiece, we will have to look elsewhere. Let's try the kitchen.

TUNING FORKS. Find a large fork, preferably of silver, but other metals will do. Place it flat on the table, with one finger pressing down as in the picture. Snap one of the prongs with your fingernail. Listen to the musical sound, made by the vibrations of the prongs. Then hold the fork upright against a wall and snap it again. You will hear the same pitch of sound, because position doesn't matter to a vibrating fork. If you had some machine to do the snapping, the fork would continue to vibrate.

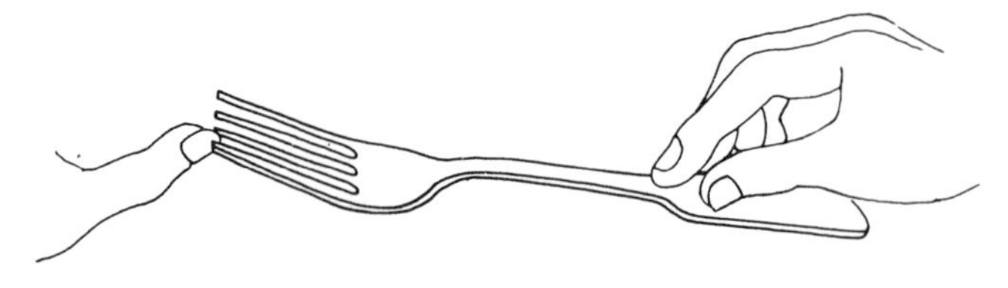

All the vibrations would take equal intervals of time. If you could count these vibrations, you would have a timekeeper that works not by a pendulum, or a hairspring, but by a fork!

And here it is. The fork is quite tiny, in order to fit into a watch case. The "snappers" are two electromagnets whose energy comes from a small electric cell. The electromagnets keep the fork vibrating for about one year, after which the cell has to be replaced. The vibrating fork controls a set of gears that move the hands of the watch. This kind of watch is accurate to about one minute per month.

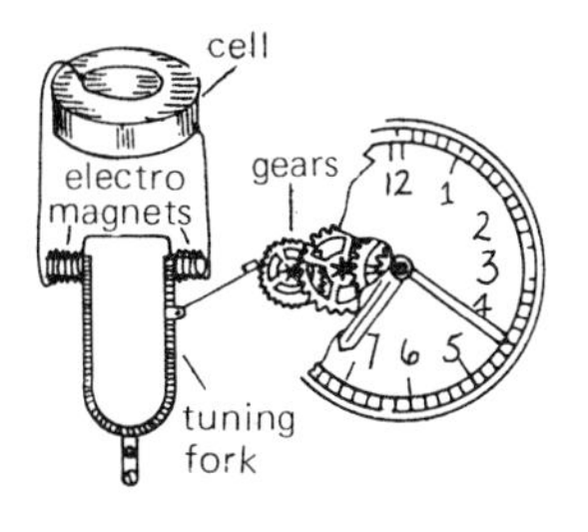

Not quite good enough. Some scientists want accuracy of a thousandth of a second per *year*. To obtain it for them, we go back to the kitchen.

CRYSTALS. Consider gelatin pudding. If you tap it with a spoon, it shakes. If you tap it on top, it shakes up and down a few times. Each shake, or vibration, takes an equal time interval (almost equal, anyway). Keep feeding in the taps and it keeps giving out the vibrations.

You can think of a couple of dozen objections to the idea of a pudding clock. But there are many substances that can be made to vibrate like a pudding. These substances are crystals. A diamond is a crystal—and expensive. A chip of quartz rock is a crystal—and cheap.

A quartz clock or watch contains a quartz crystal that vibrates 30,720 times per second. The quartz crystal is kept vibrating by a "snapper," an electric current from a small cell. The vibrations of the crystal control gears that turn and move the hands of the clock or watch. Or they control tiny switches that light up numbers to tell the time.

Good enough? Not for some purposes. For example, scientists wanted to know whether the earth rotates in exactly the same interval every day. If not, they would have to give up the old definition of a second equaling one 86,400th of a day (60 seconds × 60 minutes × 24 hours = 86,400). So, onward in the search for accuracy—not to the kitchen this time, but to the laboratory. There we examine a substance called *cesium.*

cesium clock

ATOMIC CLOCKS. Like all substances, cesium is composed of *atoms.* Atoms have a central part, the *nucleus,* made of *protons* and *neutrons.* The nucleus vibrates all the time, and it doesn't need a snapper to keep it vibrating. A cesium nucleus vibrates almost exactly 9,192,631,770 times per second. "Almost exactly" because there may be a gain or loss of one second in 30,000 years!

As you might expect, a cesium clock doesn't resemble any clock you ever saw before. Part of it is a tube eighteen feet long, and another part is an oven! It has been used for checking various motions of the earth, and other planets and stars, with some surprising results. For instance, it has revealed that in some years the earth's rotation is about a second slower or faster than others. In some seasons the daily rotation is one-fortieth of a second longer or shorter than in other seasons.

A Half Clock

Big and little, cheap or expensive, accurate or not, all clocks contain two kinds of parts. There is a time *divider* (pendulum, hairspring, fork, crystal, atom) that ticks off equal time intervals, and a time *adder* (gears and hands) that counts the time intervals and shows the results on a dial.

electric clock

Together, parts one and two add up to a whole clock.

Somewhere in your house, you probably have a half clock. It's plugged into a wall outlet. Electricity from the outlet runs a motor that turns gears and hands. The motor, the gears, and the hands all make up the second part, the time *adder.*

But where is the time *divider* that ticks off equal time intervals? It's miles away, in the powerhouse. Generators in the powerhouse make AC: *A*lternating *C*urrent. This is electric current that reverses itself, flowing first in one direction and then in the opposite direction. This forward-and-backward flow is called a *cycle.* The generator sends out sixty cycles a second, so the current is called sixty-cycle current.

Sixty cycles per second is also called 60 Hertz, or 60 Hz, in honor of Heinrich Hertz, who lived from 1857 to 1894. Hertz made important discoveries about electromagnetic waves. These waves, also called Hertzian waves, are described on pages 97–99.

The generator, spinning at exactly the right speed to produce a sixty-cycle current, is the time divider—almost. Something has to keep the generator at the right speed. A cesium clock in Washington, D.C. sends out time signals by wire and radio to many places in the country, including powerhouses. These signals are compared with the time on an electric clock run by current from the generators. If there is a difference (maybe as much as one-tenth of a second, or six cycles) the generator in the powerhouse is slowed or speeded by an engineer. This is done gradually, during the night, when it is least likely to disturb anybody who might need accurate time.

Telling Time Backward

"How long ago did a caveman build the fire that left these ashes?"

"How old is this mammoth skull?"

"When was this mummy alive?"

These are questions about materials that were once alive—wood, bone, skin. The questions can be answered because scientists have developed a method of counting backward in time. This method, called *radiocarbon dating,* is based on these facts:

1. All living things contain carbon.
2. Some of the carbon in living things changes.
3. We know how much time it takes for the carbon to change.
4. We can measure how much change has taken place.
5. Therefore, we can tell how much time has passed since the carbon first began to change.

Let's look at these statements, one at a time.

1. All living things contain carbon.

A caveman threw a log on the fire. The log was part of a plant, a tree. The tree had grown by taking in water and minerals from the soil, and *carbon* dioxide gas from the air. This process goes on in all green plants.

The mammoth ate plants.

The mummy was once a person who ate plants (for example, corn) or animals that ate plants (for example, chicken).

So, directly or indirectly, all living things contain carbon from carbon dioxide in the air.

2. Some of the carbon in living things changes.

There are two forms of carbon in carbon dioxide from the air: ordinary carbon, called *carbon 12*, and heavier carbon, called *carbon 14.* Both forms are constantly being taken from the air by plants, and are constantly being replaced by certain changes in the atmosphere.

Carbon 14 atoms behave in a special way. Each one can emit a particle of itself and turn into a nitrogen atom. In doing so, it causes one tiny flash of light to be emitted. These flashes can be detected and counted by an instrument called a *scintillation counter.* (*Scintilla* is Latin for "spark.") The atoms don't flash all at once. An atom suddenly flashes here, another atom there—and that's it for these atoms. Scientists don't know which particular atoms in a carbon 14 cluster will flash next, but they do know *how many* of them will change in a given period of time. In other words—

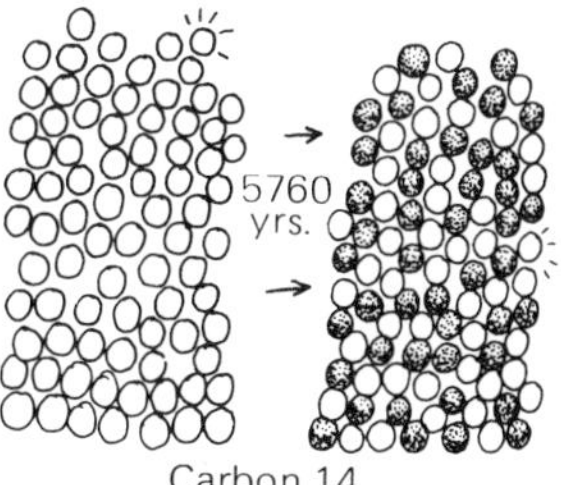

Carbon 14

3. We know how much time it takes for carbon 14 to change.

Suppose we have a cluster of newly formed carbon 14 atoms. Scientists have found that in 5,760 years half of them will have emitted their single flash. This period of time is called the half-life of carbon 14. We say that the half-life of carbon 14 is 5,760 years.

Scientists didn't have to wait 5,760 years to measure carbon 14's half-life. And you don't have to wait a whole year to measure the annual water loss from a dripping faucet, if you know how much water drips out in one minute.

4. We can measure how much change has taken place.

Suppose we have two equal-sized samples of charred wood. Sample 1 comes from a campfire we built yesterday. Sample 2 comes from an ancient but newly opened cave. Each sample is examined with a scintillation counter, with these results.

1. (Yesterday's fire): Forty-eight flashes per minute.

2. (Ancient cave): Twenty-four flashes per minute.

The ancient wood emits only half as many flashes per minute. This is because half of its carbon 14 atoms have already changed. Since the half-life of carbon 14 is 5,760 years—

5. We can tell how much time has passed since the carbon atoms first began to change.

Sample 2, from the ancient cave, is 5,760 years old. In the same way, we can date the age of other samples, older or younger. For instance, in the next half-life period, half of the *remaining* carbon 14 atoms will have emitted their flash. So the count will be twelve flashes per minute instead of twenty-four. Here is a graph of such changes.

Point A on the graph marks a sample 5,760 years old. Point B shows a sample 8,000 years old.

How about point C?

What is the flash rate of a 2,000-year-old sample?

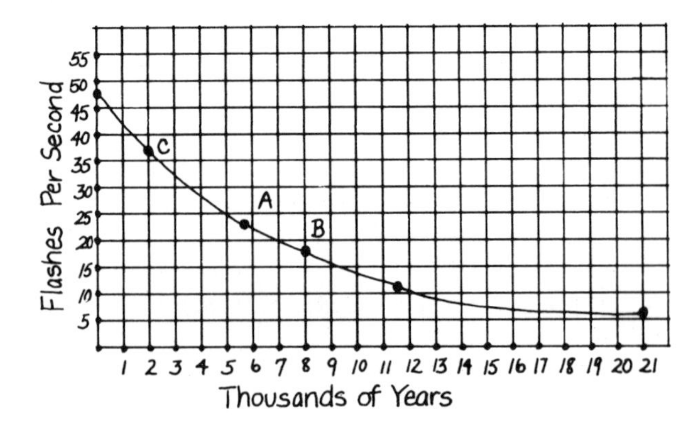

Radioactivity

The ability of carbon 14 to emit a flash, or ray, all by itself is called *natural radioactivity*. There are many other naturally radioactive elements. These can be used to date the substances in which they are found. Let's look at an example.

A scientist wants to date two samples of moon rock, from two different places on the moon. Are they of the same age, or is one older, and if so bv how much? The answer will help the scientist determine how certain features of the moon were formed.

Both samples are analyzed with a spectrometer (see page 26). Both are found to be made of the same elements. Among these elements, two are especially interesting to the scientist. These are uranium 238 and lead 206. Uranium 238 is naturally radioactive. It goes through a series of radioactive changes and finally ends up as lead 206!

We know the half-life of each change. Together, all these changes add up to about $4^1/_2$ billion years.

When both rocks were originally formed, they contained uranium 238, but no lead 206. Bit by bit the uranium 238 turned to lead 206. If the two rocks were formed at the same time, they should contain the same proportion of uranium to lead. But if one is older, it will

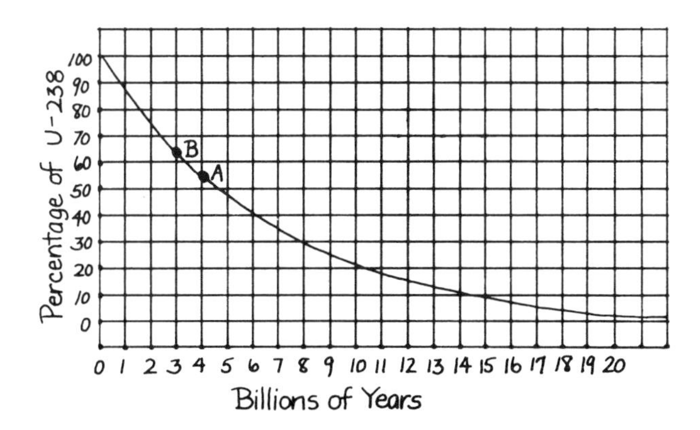

contain less uranium and more lead than the younger rock.

The scientist does a quantitative analysis on the two samples (see page 22). Then s/he refers to a graph like the one for carbon 14. This one is based on the half-life changes of uranium 238 to lead 206. Here it is: Sample A is a billion years older than Sample B!

space

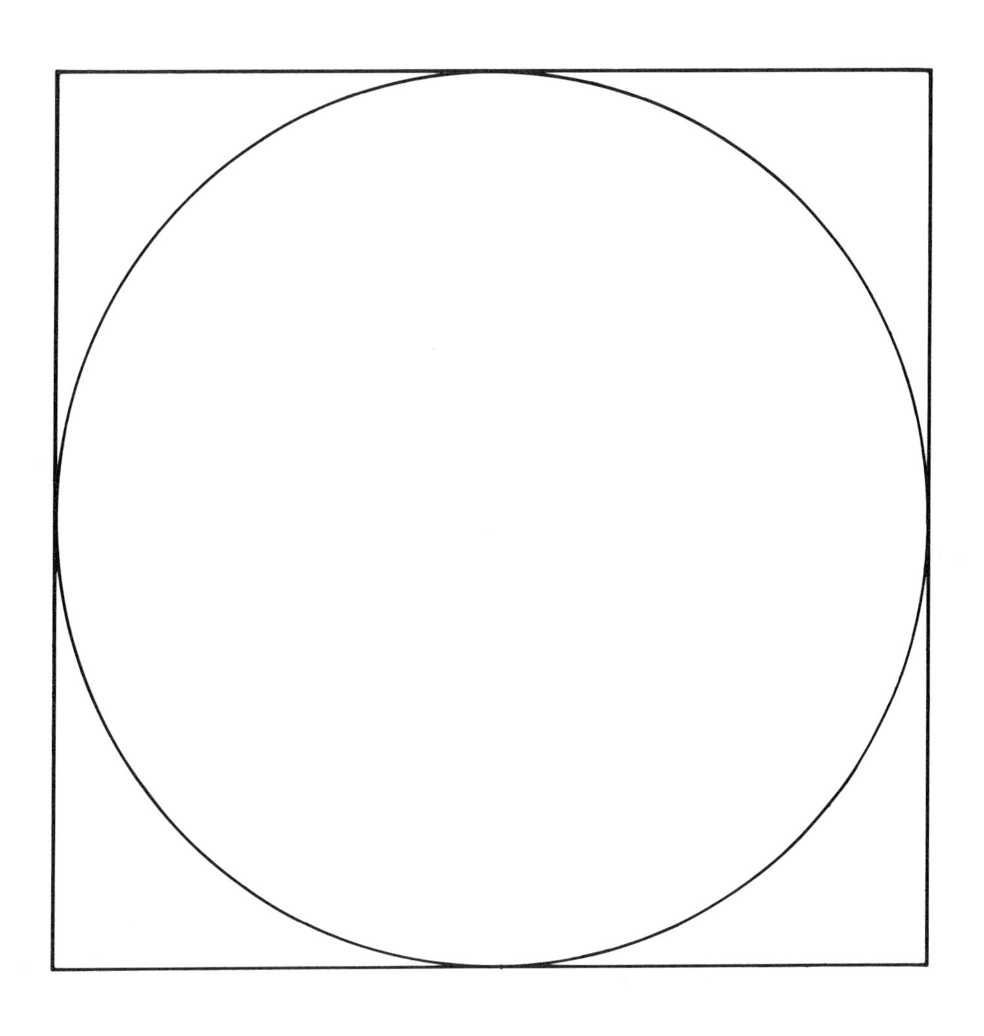

How big is that flying saucer?
How far away is it?
How fast is it going?

These are questions about a flying saucer. But they are also about *space*. "How big" could be stated another way: How much space does the flying saucer take up? "How far away" could be: How much space is there between the flying saucer and me? "How fast" could be: How much space does it move through in a minute or a second or whatever? Even the question "Is that flying saucer real?" is about space—because real things take up space and unreal things don't. A flying saucer, if it existed, would take up space that could be measured.

How Scientists Measure Space

Every scientist asks questions about space measurement, and sometimes finds answers.

An astronomer: How far away from the sun is Mars? (About 142 million miles, or 229 million kilometers.)

A physicist: How big is an atom? (About a hundred millionth of a centimeter.)

A biologist: How fast does a nerve message travel? (About 350 feet or 107 meters per second at top speed.)

Let's consider some methods of measuring space. You have used one of these methods many times.

Measuring By Direct Comparison

When you measure yourself with a ruler, such as a meter stick, you are comparing your height (unknown) with the length of the ruler (known). The ruler's length is known because it, too, has been compared with something. When the ruler manufacturer first set up his machinery, he used a metal rod whose length had been compared with a *standard meter.* This is a metal rod which is kept at the National Bureau of Standards in Washington, D.C. Other standard meters are kept in several cities in the United States and in other countries.

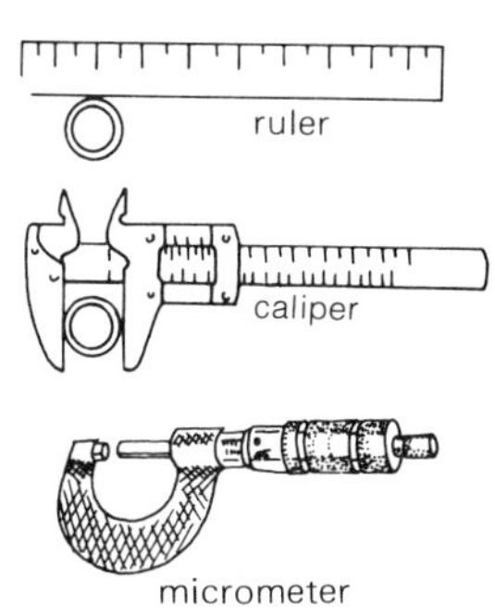

Try measuring the diameter of a pipe with a ruler and you'll see why a *caliper* is better. A caliper also works by direct comparison. But it has a moveable jaw that permits more exact measurement than a plain ruler.

This is a kind of caliper called a *micrometer* (from the Greek word meaning "small measurer"). The moveable part is a rod that is moved left or right by a screw inside the handle. Turning the handle turns the screw which moves the rod. The number of turns is counted on a scale, marked off in fractions of inches or millimeters.

Micrometers are the most precise of the instruments that work by direct, physical measurement. A micrometer can easily measure the thickness of a sheet of tissue paper or the difference between a thin hair on your arm and a thicker hair on your scalp.

Measurement By Reflection

Bats can fly around in a pitch-dark room crisscrossed with wires, without striking a single one. Bats emit high-pitched squeaks which are reflected by the wires. They can sense the space between themselves and the wires by sound reflection, echos. If the pause is short, the wire is close, and a long pause means the wire is farther away. Of course these short and long pauses are both quite short to human senses. Sound doesn't take very long to travel from the bat to a wire a foot away and back to the bat's ears. Sound travels about 1,100 feet per second.

Suppose you uttered a good loud yell at a distant cliff and heard the echo two seconds later. The sound took one second to reach the cliff and another second to return to your ears. You could then estimate the distance between the cliff and yourself to be 1,100 feet.

This system of measurement by reflection, or echo, is used in many ways by navigators, pilots, fishermen, geologists and astronauts. Distance is measured by the reflection of sound waves in air, sound waves in water, radio waves in air and in empty space. Let's look at a few examples:

Sound Waves In Water

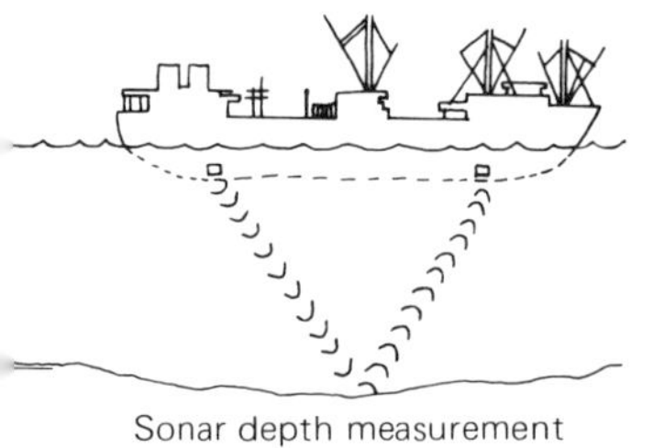
Sonar depth measurement

An oceanographer wants to locate an undersea mountain. A ship's pilot wants to know the depth of water in a harbor. A fisherman wants to locate a school of fish. A geologist wants to locate an oil field deep under the ocean floor.

All four can get their answers through the use of a system of instruments called sonar. The name was formed from *SO*und *NA*vigation and *R*anging.

A sonar system has three main parts. First, there is a *sound-maker* attached to the bottom of the ship. The sound-maker sends short pulses of sound—beeps—down through the water. Sound travels through water at about one mile per second. The sound pulses strike the bottom and echo upward to a *sound receiver*, also at the bottom of the ship, which receives the reflected sounds and converts them into electric signals that go into a *timer, calculator and recorder.* The timer measures the time interval between the outgoing beep and the returning echo. The calculator converts this time interval into distance. For instance, if the ocean is a mile deep, the sound will take one second to reach bottom and another second to echo back to the ship. The calculator computes this two-second interval as a depth of one mile. This is shown on a dial or is marked on a graph. To make a map of the ocean floor, the ship moves along at a steady speed while the sonar keeps sending signal after signal.

Now what about the fisherman? Sound waves are reflected by solids, such as sand or rocks at the bottom of the sea. But fish are solids, too. Schools of fish reflect enough sound energy to work the sonar receiver. But there is a difference between the steady reflection from the sea bed and the scattered reflection from fish.

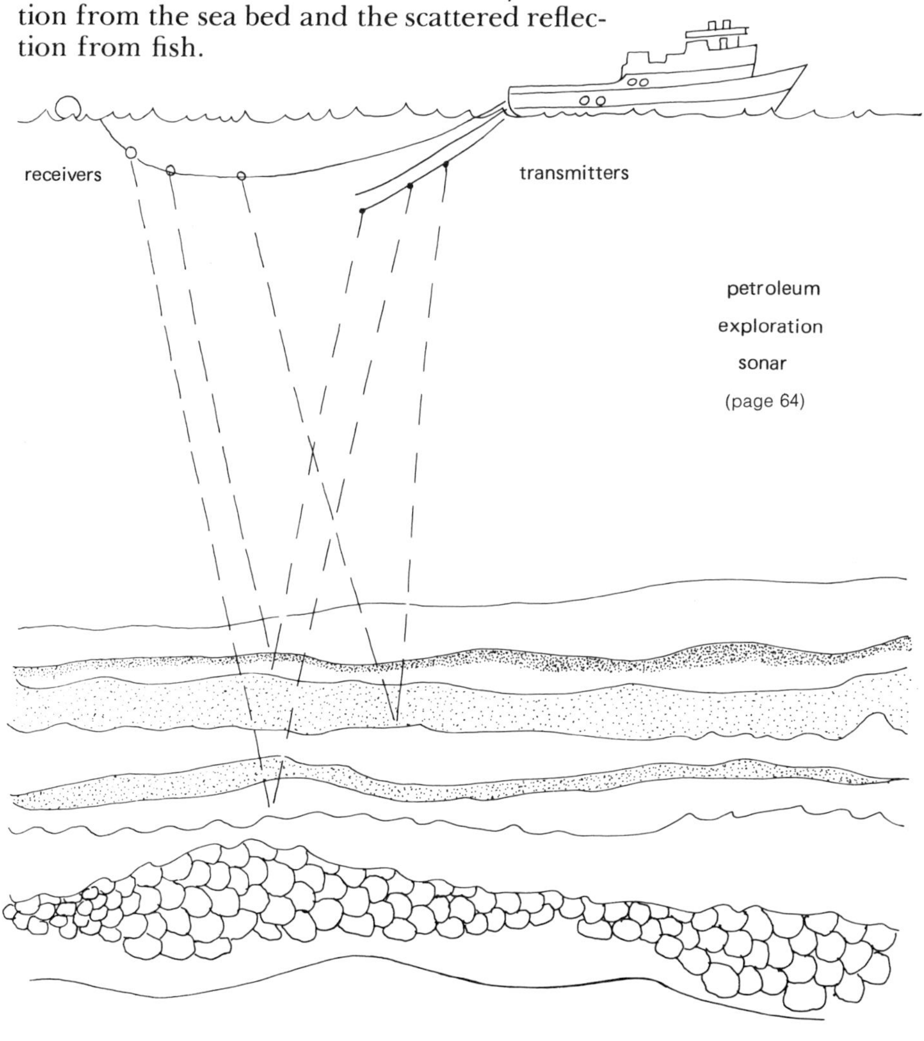

receivers

transmitters

petroleum exploration sonar (page 64)

The geologist seeking oil uses a large and complicated sonar. The sound-maker is an air machine that shoots out sharp "whump" sounds down to the sea bed. The sounds are powerful enough to travel still further to rock layers thousands of feet below the sandy floor. The echos from the rock layers, and a separate echo from the sandy bottom, are both reflected to the sound receivers.

There are several thousand sound receivers attached to long cables trailing behind the ship. The signals are recorded on a chart and fed into a computer. The results help geologists to determine the thickness and shape of the various rock layers, which will indicate to them whether oil is likely to be present or not.

Radio Waves Through Air And Space

"What is my altitude?" asks an airplane pilot. In the olden days of aviation the answer came from the *pressure altimeter* on the instrument board. This altimeter worked by air pressure. The higher the plane flew, the thinner the air, and the weaker the air pressure. The pressure was shown by a pointer on a dial, in the form of "altitude above sea level."

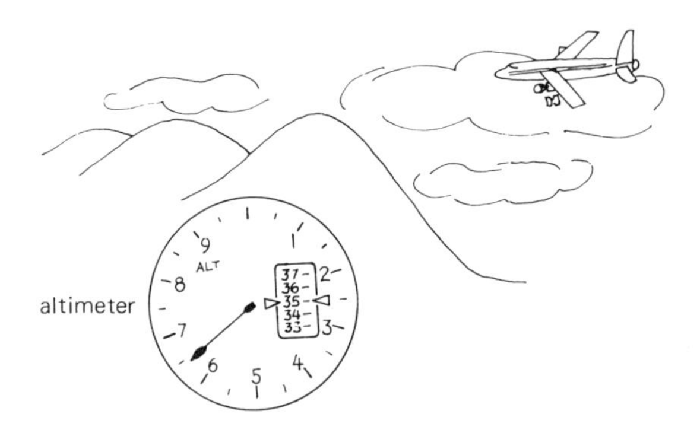

But an airplane doesn't always fly above the sea, nor above flat land at sea level. The pilot also wants to know the actual altitude above the land or water.

The answer comes from a system of instruments called radar, which stands for *RA*dio *D*etection *A*nd *R*anging. Like sonar, this system has three main parts:

A *transmitter* sends pulses of radio waves down from the airplane. The waves strike the ground or water and reflect back to a *receiver* in the airplane, which receives the reflected pulses and converts them into electric signals that are fed into a *timer, calculator* and *dial.* The timer measures the time interval between the outgoing radio wave pulse and the returning echo. The calculator converts this time interval into distance, and shows it on the dial or marks it on a chart.

When a radar transmitter is pointed downward, it measures altitude above the ground or water. Pointed forward, it measures the distance to a mountain ahead or to a city hidden in fog. A special kind of radar can even detect and range the water droplets in distant cloud formations, to help the pilot decide whether to change his course. This is what a hurricane looks like on the screen of such a radar:

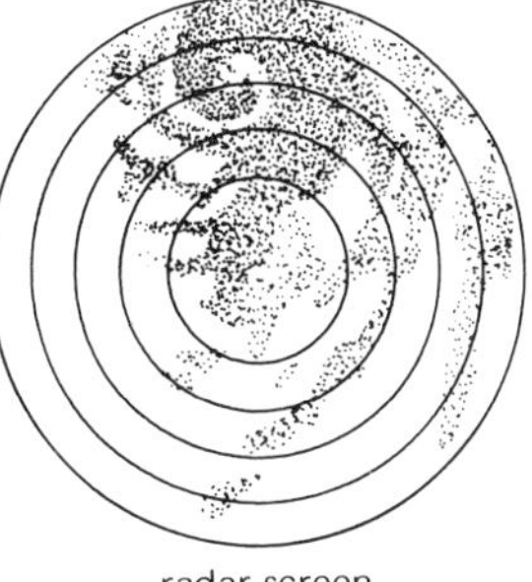

radar screen

There seems to be no limit to the usefulness of radar. Distance above the ground, distance to hills ahead, distance to weather disturbances—how about distances to outer space?

Measuring the distance to the moon is an easy job for radar. The average distance is 238,857 miles or 384,321 kilometers. The moon's orbit is not a circle but an *ellipse.* The distance varies from day to day and even from moment to moment. But radar can keep track of these variations.

The first astronauts to land on the moon carried an accurate map with them. It showed every detail of the region where they were to

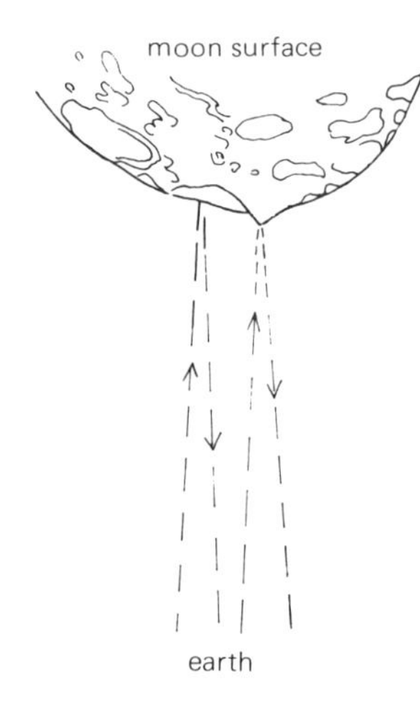

land. The map was much more detailed than an ordinary auto road map. It showed the height of hills, the depth of craters, the width of rills and valleys. All these sizes and shapes, these accurate dimensions, had been mapped from earth and from moon satellites with the aid of radar.

The distance from earth to moon depends on which part of the moon is reflecting the radar waves. The peak of a moon mountain, in this position, is closer to earth than the plain next to it. The radar waves are aimed first at the mountain peak and then at the nearby plain. Subtracting one distance from the other gives the height of the mountain. Section by section the radar system measured and mapped the entire side of the moon facing the earth.

Microscopes

From your window you can see the surface of the moon. Astronauts on the moon saw it much better. Standing on the shore you can see the sandiness of the beach. A sand flea on the sand can see it much better.

You could say the closer you look the better you see.

But there are two problems. You can't always come closer, and if you come too close your eyes can't focus; you see a fuzzy image.

For the first problem we use a *telescope* and for the second we use a *microscope.*

First let's investigate how we see without the help of an instrument. This is a diagram of an eye looking at a candle. Light from the flame

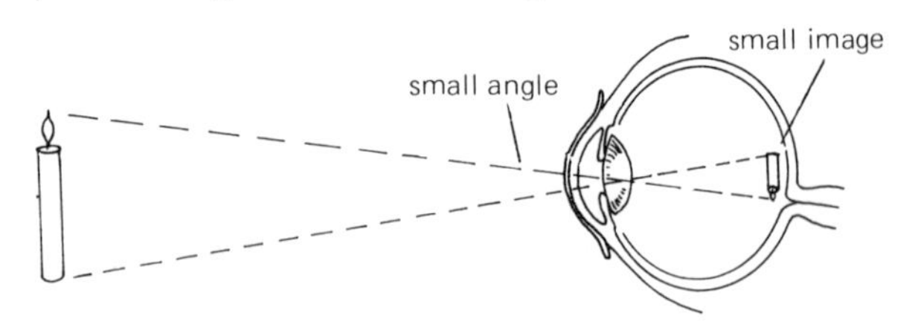

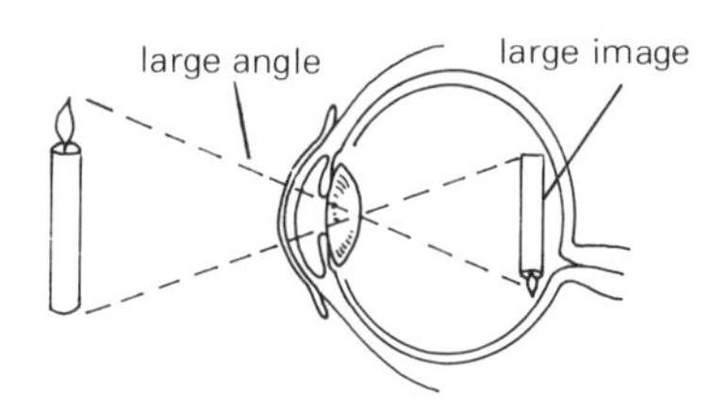

and light reflected from the candle enters the eye and forms an image on the back of the eye, the *retina.*

If the candle is brought closer, the light enters the eye at a larger angle and forms a larger image.

Still closer, we get a still larger angle and a still larger image.

However, there is a limit to how close we can come. Try it with the words you are reading. When you come too close they become fuzzy, out of focus. To see clearly we need help. Here's the earliest form of help: a caveman's magnifying instrument—a drop of water. (Not that we know whether the caveman actually used drops of water as magnifiers, but you can.)

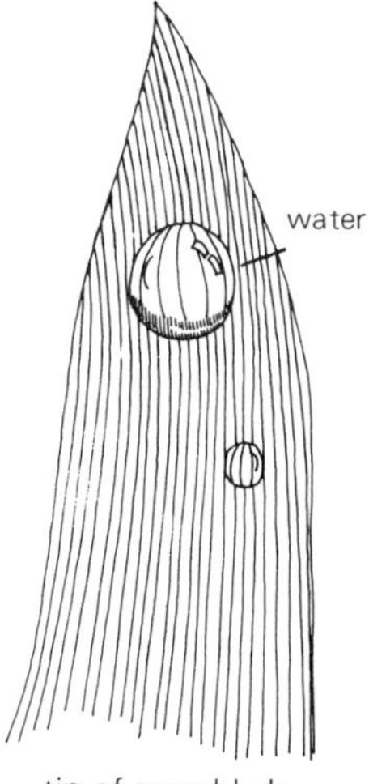

tip of grass blade

With a matchstick or eyedropper, put one drop of water on a leaf. Observe the magnifying effect. The water bends the light rays so they enter your eyes at a larger angle and produce a larger image on your retina.

You can use your magnifier on other objects too. Lay a bit of cellophane or clear plastic wrap on a newspaper, then put a drop of water on it. The newsprint looks larger—until the water evaporates.

A scientific instrument that evaporates in a minute isn't too desirable. So we use glass instead of water. This watch repairman is using a magnifying glass to get a magnified view of the insides of a watch. The magnification is accomplished in two ways: (1) The lens can form a clear image from a closer position

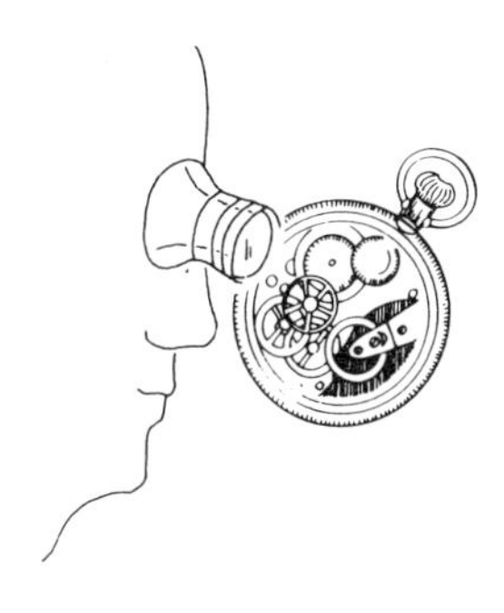

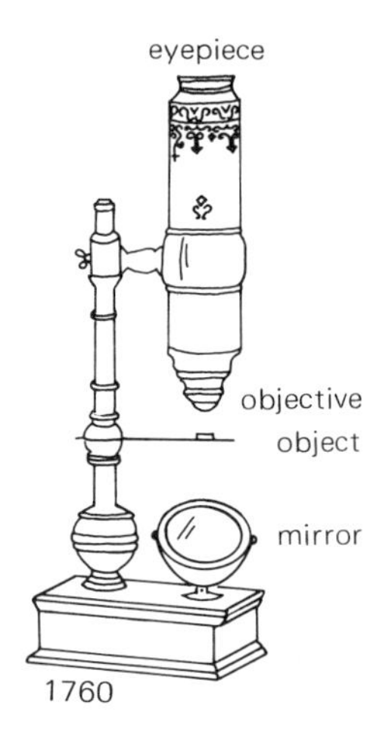

1760

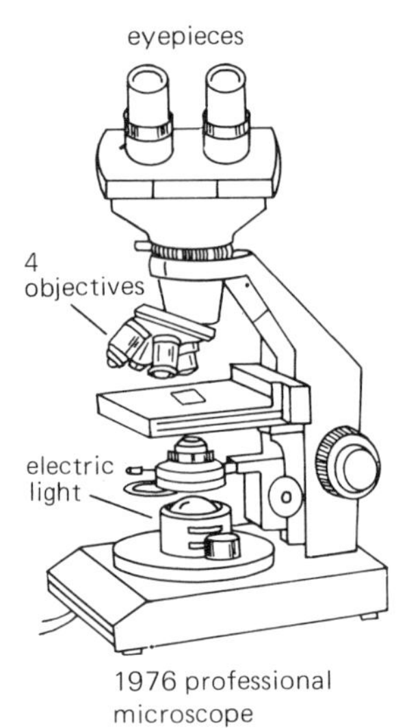

1976 professional
microscope

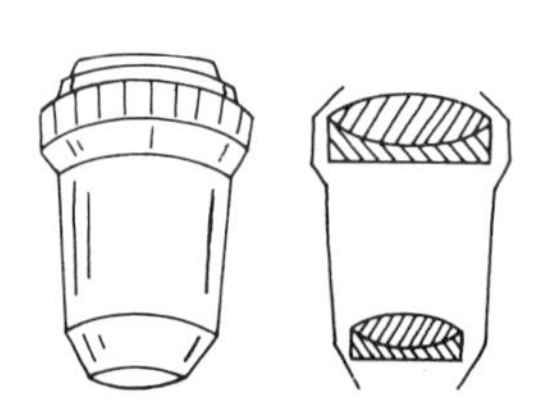

than the eye alone can (closer equals larger); (2) The lens is shaped so that it bends the light rays to form a larger angle (larger angle equals larger image).

Even so, there's a limit to how much clear magnification you can get with a single magnifying glass. So let's use two magnifying glasses. Such an instrument is called a *compound microscope.*

Actually a good compound microscope has two or more *groups* of lenses. The lower group is called the *objective* because it is closer to the object being examined. The objective forms a magnified image of the object. The upper group is called the *eyepiece* because it is close to the eye. The eyepiece magnifies the magnified image into a still larger image. The total magnification is the product of the two. For example, an objective that magnifies fifty times and an eyepiece with a magnifying power of ten times, together produce a magnification of 500 times, or 500X. The highest power of this type of microscope is 2,000X.

Why two *groups* of lenses? Because single lenses suffer from all kinds of handicaps. For instance, there is a handicap called *chromatic aberration.* This can be loosely translated as off-coloredness. A single lens can produce a sharp image of only one color at a time; other colors are unclear. If you form an image of an American flag so that the red stripes are sharp, the blue star field will be out of focus, and vice versa. Every single lens suffers from chromatic aberration. You need a group of two or more lenses to correct this condition.

There are still other conditions to worry about that require still more lenses. So you can understand why this objective lens group has four separate lenses in it, and some have as many as seven. The eyepiece too has its own set of problems with its own set of solutions.

Altogether, to design and construct a good compound microscope calls for lots of scientific knowledge and skill.

RESOLVING POWER. We have been using the words "sharp" and "unclear" in a general sort of way. But that's a rather unclear way to describe a lens. A scientist doesn't go to a microscope store and ask for a "very sharp" instrument. Nor does he reject another because its image is "not clear enough." Instead he discusses *resolving power*, and so will we.

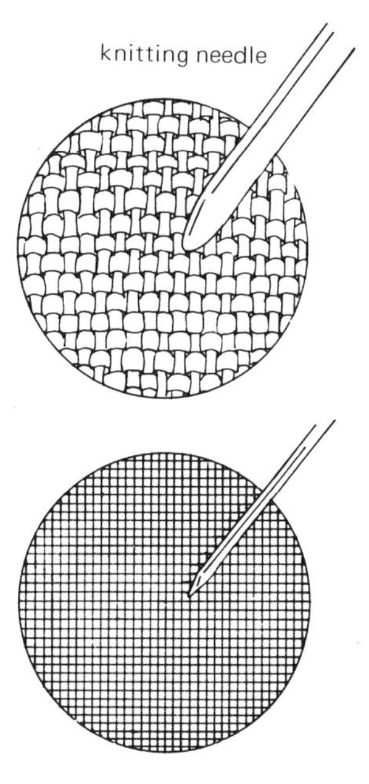

Imagine that you're trying to count the threads in a loosely woven rug. You can make your job easier by running the point of a knitting needle across the threads, feeling the bumps as you go. A fairly fat knitting needle will do, because the threads themselves are comparatively thick.

The next job is to count the threads in a tightly woven silk scarf. The knitting needle won't do—it's too thick to separate, or *resolve* the cloth into its fine thin silk threads. You need something with a higher resolving power such as a sewing needle. The point of the needle is thin enough to resolve the silken weave into its separate threads.

Now let's see what resolving power has to do with microscopes. On page 70 is a magnified view of a tiny water plant called a *diatom.* Those tiny circles on it are a hundred thousandth of an inch wide! Yet you can clearly see each circle, because the photograph was taken through a microscope of high resolving power. Compare it with photograph B, taken with a microscope of low resolving power. No problem guessing which microscope costs twelve hundred dollars and which twenty dollars!

To get back to the needle and cloth comparison. You can figure that the limit of resolving power of the needle is the thickness of one thread. The needle point can't be thicker than the threads it's counting, or it won't be able to separate (resolve) them. The fat knitting needle was all right for the woolen rug, but not for the silk scarf.

The same thing is true for microscopes.

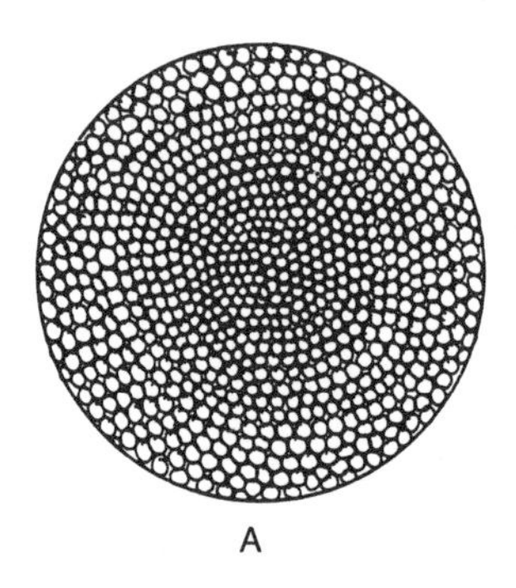

A

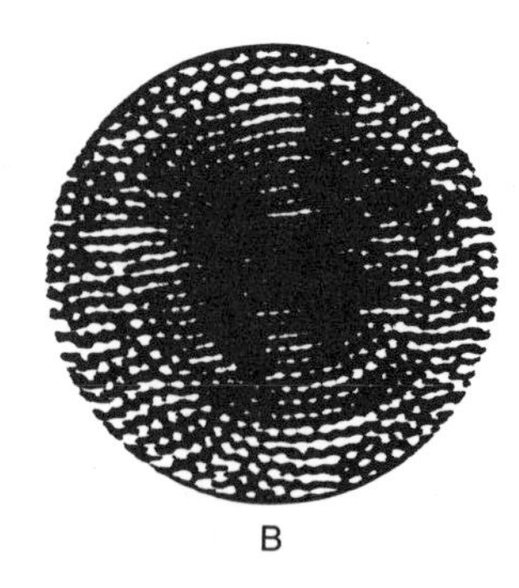

B

Compare the diatom dots to threads that have to be separated and counted. The "needle" is the rays of light shining through the diatom into the lenses of the microscope. If the lenses have high resolving power, they can handle those separate rays and keep them separate; we get a picture like (A). With lenses of low resolving power, the light rays become slightly overlapped, and (B) is the result.

Some objects, such as an influenza virus, are thinner than light rays! It's like having cloth with such a fine weave that even the finest steel needle point is too thick. Then we have to look for other metal that can be made into a still finer point. With microscopes, we have to look for rays that are thinner than light rays.

ELECTRON MICROSCOPES. Beams of electrons are super-thin. They permit a much higher resolving power than light rays, but there are two problems.

Electron beams are invisible to human eyes.

Therefore, the beams have to be aimed at a screen, like a small TV picture tube. The image on the screen is visible, and can be photographed, too.

Electron beams can zip right through glass, without being bent. Therefore, glass lenses won't do for focusing an image. But electromagnets can. Round electromagnets work like the groups of glass lenses in an optical microscope.

You can see the action of a magnet on an electron beam if you have a black and white television. Turn on a picture, hold the magnet about an inch from the screen and move it slowly from right to left. Watch the picture wiggle and follow the magnet. Your TV picture is made by electron beams at the rear of the picture tube striking the screen in the front.

Electron microscopes are much more costly than light (optical) microscopes. They are much more difficult to operate and keep in condition, but the best optical microscope can give an image of good resolving power with a magnification of 2,000X compared to an electron microscope's 200,000X!

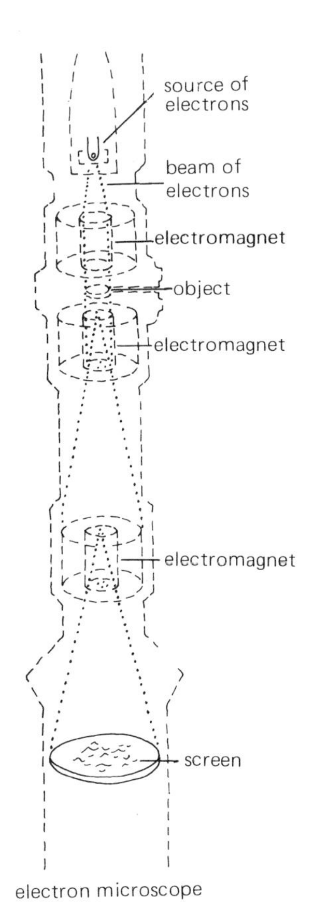

electron microscope

Telescopes

Telescopes and microscopes work in very much the same way. They increase the angle of light rays so as to form a larger image on your retina.

You can assemble a miniature telescope in a few minutes. You will need a small square of wax paper, a ruler and two *different* magnifying lenses. Their difference has to be in focal length. Here's how to tell:

Hold one of the lenses near the window and form an image of the outside view on a sheet of paper. Note the distance from the lens to the paper. Then do the same with the other

lens. The one that makes an image at a greater distance has the longer focal length. This will be the objective lens. Mark it O. The other will be the eyepiece. Mark it E.

Darken the room except for one shade half drawn. Stand a thick book near the window. Support lens O by its handle in the book. Hold the wax paper so that a clear image of the outside scene is focused on it. It's a small image. How can we make it bigger?

Hold lens E so that it acts as a magnifier of the image on the wax paper. Now you have a larger image.

But real telescopes don't have wax paper inside. Take away the wax paper—you'll still see an enlarged image.

Objective lens and eyepiece lens—these are the principal parts of most telescopes, as they are of most microscopes. And as with microscopes, they magnify in the same way, by increasing the angle of the light waves entering your eye. They do this by bending (*refracting*) light rays, so they are called *refracting telescopes.*

There are differences, of course, between microscopes and telescopes. One difference is the matter of *light gathering.* When you're examining an insect wing, you can shine a light on it to form a bright image, if you wish. But you have no such privilege with a telescope. The faraway zebra, planet, or star is lit up in its own way. There is nothing you can do to increase the illumination of the object. But you can do something about increasing the amount of illumination caught by your telescope. Just increase the diameter of the objective lens. The bigger the lens, the more light it catches, and the brighter the image—to a limit.

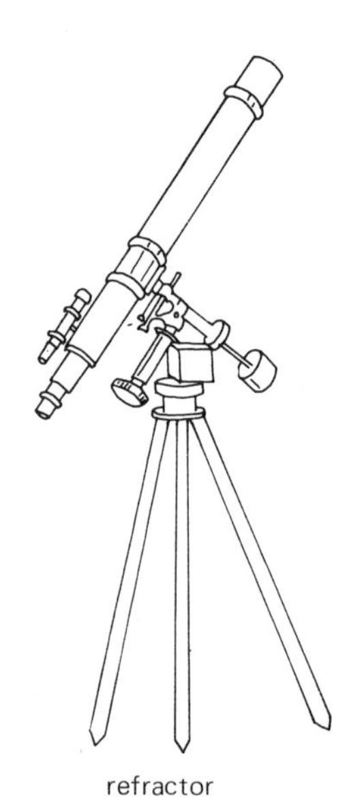
refractor

The limit is weight. You can't make a glass lens too big because it gets too heavy. The lens is supported only at its rim. Its own weight causes a lens to bend slightly. A thousandth of an inch will confuse the image. To collect the most possible light from distant dim stars, we use a *curved mirror* to focus an image instead of a transparent glass lens. A mirror can be built much larger and heavier than a lens, because it can be completely supported from underneath. A large heavy mirror can collect lots of light. It can make bright images for the astronomer to look at through a magnifying eyepiece or to photograph on a film.

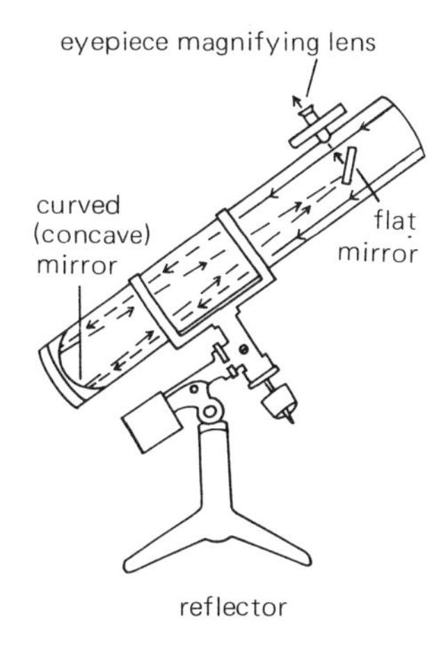

reflector

Perhaps you haven't thought of a curved mirror as working like a lens. But you can see it for yourself. Pull a window shade halfway down. Hold a magnifying shaving mirror near the window. Move the mirror back or forth, right or left, until you get an image on the wall next to the window. It's a moderate size image—how would you magnify it?

With a magnifier, of course. Hold the mirror as in the picture, while someone else examines the image on the wall with a magnifying glass. There's the magnified view of the outside scene. Telescopes using mirrors are called *reflecting telescopes* or *reflectors*. The largest telescope in the world, at Mount Palomar, California, has a mirror 200 inches in diameter (5.08 meters).

energy

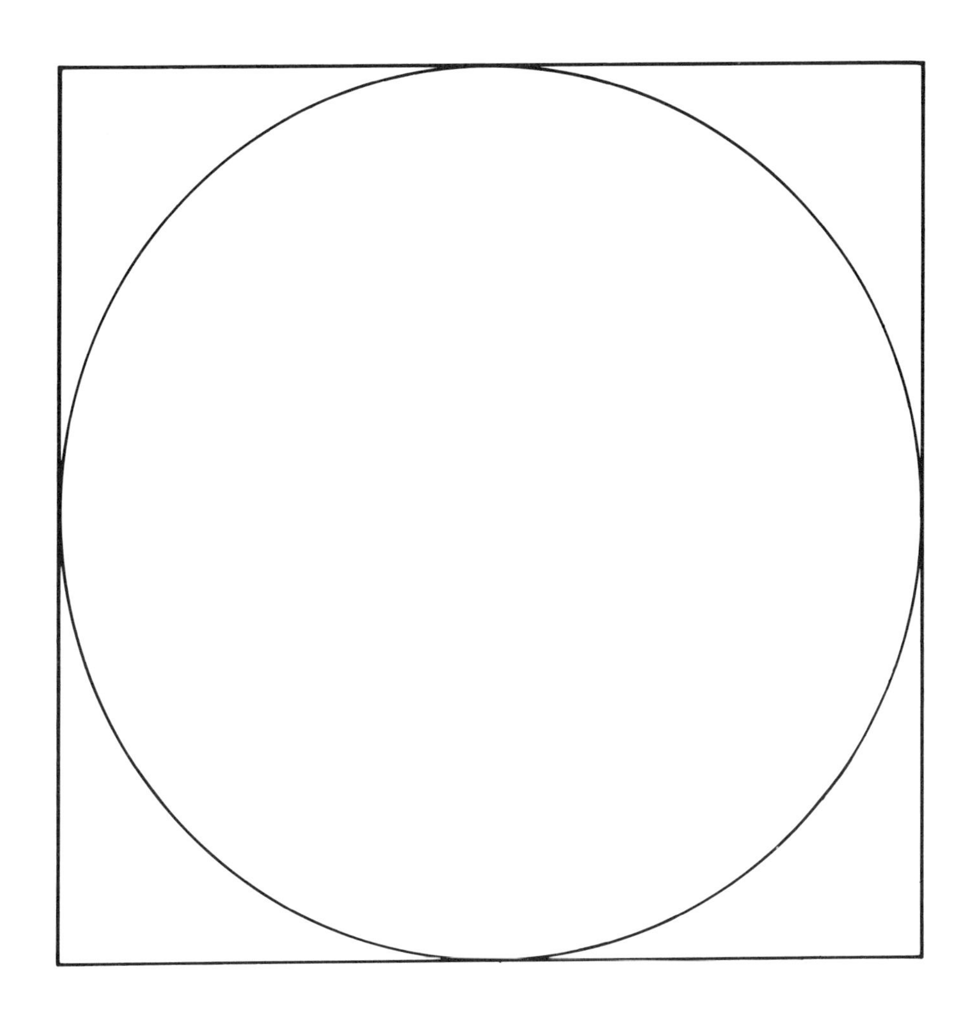

When you shoot an arrow, what keeps it going after it leaves the bow?

In the fourth century B.C., Aristotle of Greece gave this answer: "The arrowhead compresses the air in front of it. The compressed air flows around to the tail of the arrow and pushes it forward."

Today scientists can test such an answer. Set up an arrow-shooting machine inside a large sealed chamber. Pump out the air and let the machine shoot an arrow. Does the arrow fly forward or does it drop to the floor of the chamber? It flies forward.

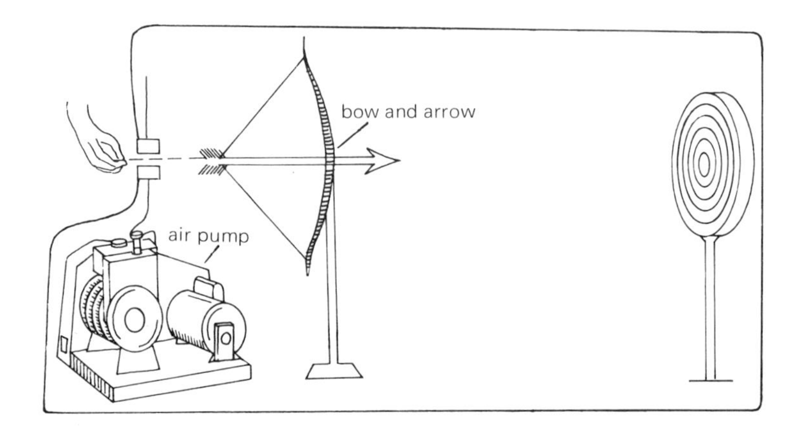

So we don't need air to keep the arrow moving. We have eliminated one possible explanation for its motion.

Questions about motion are questions about *energy*. All the happenings of the world—arrows flying, winds blowing, people sneezing, volcanoes erupting, planets spinning—all these things happen because energy makes them happen. And in one way or another, every scientist's work has to do with energy. A plant physiologist analyzes the chemical energy in a leaf. An astrophysicist analyzes the light energy in a star. Hundreds of different fields of science, and all are involved with energy.

With so many different kinds of things happening, you might expect to find many different kinds of energy. Actually, there seem to be only half a dozen different kinds, or *forms,* in two different *states.*

Potential Energy and Kinetic Energy

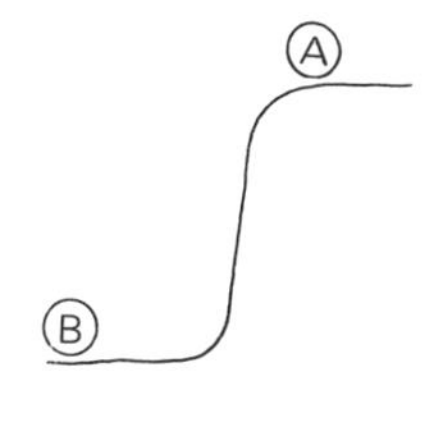

Ball A, which is at the top of the hill, is in a different state of energy from Ball B, because of its position. Ball A is *able to* roll downhill. The Latin word for "able to" gives us the word *potential.* Ball A has potential energy. Ball B, at the bottom of the hill, is unable to move by itself. It has no potential energy.

If you tip Ball A over the edge, it will start to roll. The potential (able to move) energy will become *kinetic* energy (from the Greek word for "motion").

Here are some other examples of potential and kinetic energy.

Potential: able to move.

Kinetic: moving.

POTENTIAL	KINETIC
Raised hammer	Moving hammer
Radio off	Radio on
Standing truck	Moving truck
Unlit candle	Lit candle
Unlit flash-light	Lit flash-light

It's clear enough that a hammer can move and so can a truck. But what about a flashlight, or a radio, or a candle? What's moving in each of those? The answer to that question leads us to the six *forms* of energy. Each form is determined by the answer to the question: "What is moving?"

Mechanical Energy

Let's begin with the most obvious one, the hammer. The whole hammer is moving. The hammer is a huge mass of molecules that make up the wooden handle and the steel head. All the molecules moving together possess *mechanical* energy. When the head of the hammer slams against the nail, the mechanical energy of the hammer is given to the nail. The nail moves. In the same way, a rolling truck is an example of mechanical energy, because the whole truck is moving.

Scientists and engineers have all kinds of instruments and methods for measuring mechanical energy. Let's look at one.

MEASURING THRUST. Suppose a mechanical engineer has invented a new type of engine for airplanes. One immediate and important question is "How powerful is it?" How much push, or *thrust*, can it give to the airplane? The engineer measures thrust with an instrument called a *dynamometer* (from the Greek for "power measure"). It looks quite complicated, but the basic idea is quite simple. You can construct a makeshift one yourself, to measure the thrust of a model airplane. Here it is.

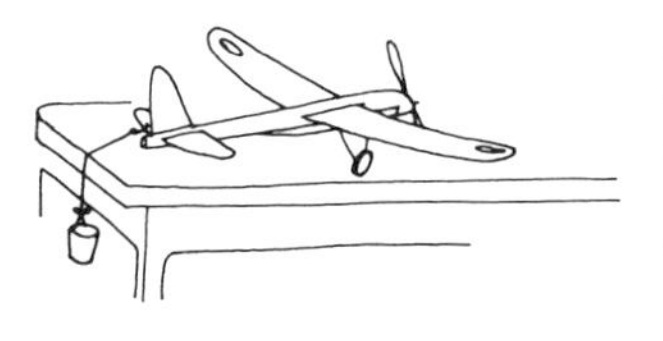

When you wind up a rubberband airplane motor and let it go, it spins the propeller. This thrusts the air backward and pushes the airplane forward. If it's free, it flies off with a whoosh. But suppose you add a load, such as a paper cup on a string. Now the takeoff is less speedy. Put a bit of sand in the cup—still slower takeoff. Bit by bit as you add sand to the load, you will reach a point where the load just balances the thrust. That is, the thrust is unable to pull the airplane forward—*and* the load is unable to pull the airplane backward. At this point we weigh the load. Suppose it weighs 1½ ounces, then the thrust of the rubberband motor is 1½ ounces.

How about the big stuff—the engines that send a transport airplane on its way? Each of the three engines of a Lockheed L-1011 has a thrust of 42,000 pounds. Each of the five engines that sent the first space ship (Apollo XI) on its way to the moon had a thrust of 1,530,770 pounds. The total thrust was five times as much: 7,653,850 pounds. The space ship, passengers, and fuel together weighed 5,022,674 pounds, so there was enough extra thrust to shoot the ship into space. Suppose the thrust had been five million pounds—no go.

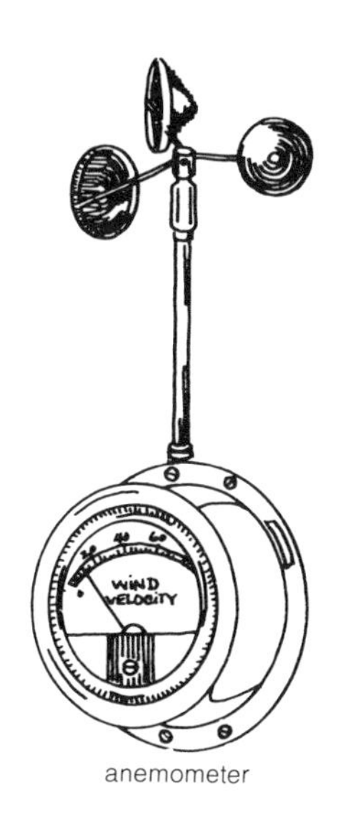

anemometer

MECHANICAL ENERGY IN A FLUID. Lots of molecules moving together in the same direction—that's the definition of mechanical energy. It applies to solid objects like hammers and space ships and to fluids like rivers and wind. Scientists and engineers do all kinds of measurements on the mechanical energy in fluids.

Here for example is an *anemometer*, a familiar instrument used by meteorologists. The wheel of half-cups is turned by a stream of moving air—the wind. The faster the stream, the faster the wheel turns. The speed is shown on a dial on the base of the instrument.

This water wheel was designed to be turned by the mechanical energy in a stream of falling water. The wheel, in turn, was to drive a machine for grinding corn. The sketch was drawn in the early sixteenth century by Leonardo da Vinci—inventor, scientist, engineer, poet, painter.

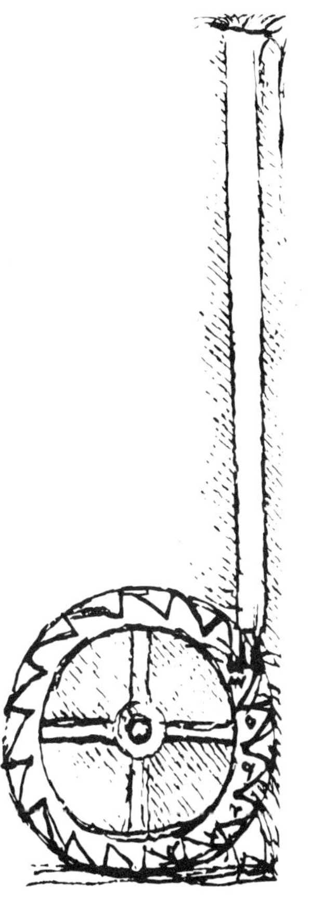

In Leonardo's design, water flows from the top of a dam, falling through a tube and slamming against the blades of the water wheel. Is this the best arrangement for getting the most mechanical energy out of the falling water? Leonardo wondered about that. He drew a sketch of four possibilities (page 80). Beginning at the top, these are labeled a, b, c and d. (You will recognize only the "a" because Leonardo often wrote backward, and sometimes upside down!) Here are his thoughts about the four possibilities, translated from the Italian:

"Which of these four waterfalls has more

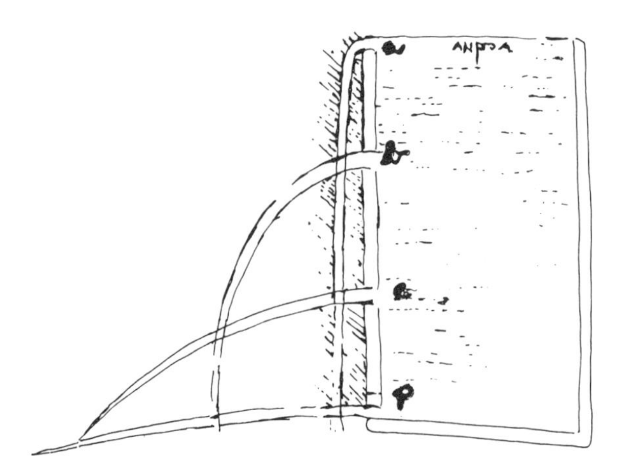

percussion and power in order to turn a wheel: fall a or b, c or d? I have not yet experimented, but it seems to me that they must have the same power, considering that a, even if it descends from a great height, has no other water chasing it, as has d, which bears upon itself the whole height of the thrusting water. Now, if fall d has a great percussion, it has not the weight of the fall a. And the same is true for b and c. Consequently, where the force of percussion is lacking, it is compensated by the weight of the waterfall."

Leonardo concluded that all four waterfalls would deliver the same amount of "power" (mechanical energy). Modern measurements with precision instruments have proven Leonardo correct.

Here is another sketch by Leonardo, done in 1493. It, too, has to do with mechanical energy. As far as is known, the machine was never built, but it certainly looks familiar. The first chain-driven bicycle was built by an Englishman, James Starley, almost four hundred years after Leonardo's sketch. Who should get the credit for the invention? No easy answer!

MECHANICAL ENERGY FOR YOUR EARS. When you pluck a guitar string, your finger puts mechanical energy into the string. The string gives back the energy by shaking back and forth—vibrating. The molecules of the string, vibrating together, give off a special kind of mechanical energy called *sound.* The vibrating string, in turn, causes the air to vibrate in the same special way. The air vibrations cause vibrations in your ear drums—and you hear.

Scientists make use of a special kind of sound called *ultrasound.* This is very high-pitched, above 20,000 vibrations per second. You can't hear ultrasound (the highest pitch you can hear is about 18,000 vibrations per second). But it possesses considerable mechanical energy. Its principal use is to split up plant and animal cells without smashing them.

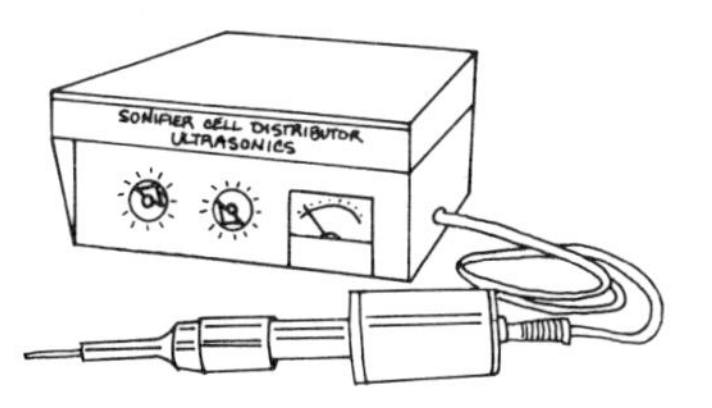

The tip of the ultrasound apparatus is placed in a liquid containing the cells, and the sound generator is turned on. Within a few minutes the cells have disintegrated into their various substances, to be further analyzed.

In mechanical energy a whole group of molecules moves together. Now let's look at the movement of separate molecules moving in many different directions. This form of energy is called—

Heat Energy

When you think of heat, you think of fire. Can there be heat without fire? Try this:

Place your palm against your forehead. They probably feel equally warm. Then rub your hands briskly together for about twenty seconds. Again place your palm against your forehead. Which feels warmer?

The molecules of your palm are in constant motion. They vibrate back and forth, side to side, up and down. The molecules of your forehead are also in constant motion. And so, for that matter, are the molecules of almost everything in the entire universe. This helter-skelter vibration of molecules is the form of energy called *heat.*

When you rubbed your palms together, you increased the speed of vibration. More speed, more heat energy. All heat energy comes from the vibration of molecules.

Scientists have devised lots of instruments for exploring heat energy. Those instruments that measure the *rate* of vibration—temperature—of molecules are called *thermometers.*

LIQUID THERMOMETERS. Most thermometers contain a liquid, either mercury (a silvery metal) or alcohol (a clear liquid colored with red or blue dye). A rise in temperature causes the molecules of the liquid to vibrate faster. They move and bounce more violently and this causes them to spread apart.

It's like a line of people standing shoulder to shoulder. The line is a certain length. But if the people start to move around, they take up more space and the line gets longer. The faster they move, the longer the line. With a rise in temperature the liquid in the thermometer rises to a higher number. With a drop in temperature (slower moving molecules) the line contracts down to the lower numbers.

In some situations liquid thermometers won't do. For instance, liquid mercury turns solid at about minus 39°C. (about minus 38°F.), so you couldn't use a mercury thermometer in many cold situations. On the other hand, alcohol boils at about 65°C. (149°F.), so it won't do in many hot situations. You have to choose the right liquid thermometer for the right use—or you can use a solid thermometer.

SOLID THERMOMETERS. Solids, too, expand and contract with changes in temperature. Each kind of solid has its own rate of expansion and contraction. You can test this for yourself with a three-inch by one-inch strip of silvery paper. Some candy wrappers, cigarette packages and fancy wrapping paper is made of this, which consists of aluminum cemented to white paper. So the strip really is two different solids: aluminum and paper.

Begin by smoothing the strip flat and straight. Now warm the strip. If the paper and aluminum expand at the same rate the strip will remain straight. But if one material expands faster than the other, the strip will bend. It will bend toward the slower-expanding material. Try it by holding the strip near an electric light or over a warm stove or a radiator. Then let it cool.

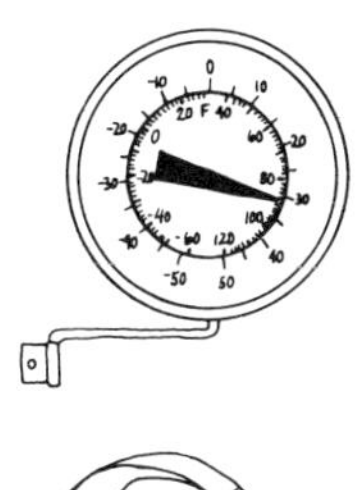

Ordinary solid thermometers have a dial and pointer. Instead of a metal-and-paper strip they have a strip made of two different metals, and are called *bimetallic* thermometers. The strip is bent into a coil shape in order to take up less room.

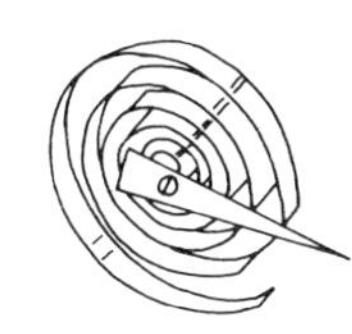

However, there's a limit to the use of bimetallic strips. If they are heated red hot, they lose the ability to return to their original

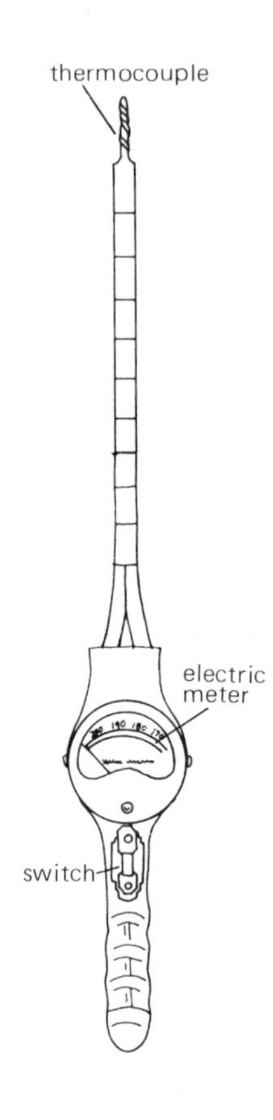

shape. So, for measuring higher temperatures, we use *pyrometers* (from the Greek "fire-measurer").

One type of pyrometer contains two wires, each made of a different metal, twisted together at one end. Such a pair is called a *thermocouple* ("thermo" from the Greek for heat; "couple" from the Latin for bond or link). When a thermocouple is heated at the twisted end, an electric current begins to flow from one wire to the other. The higher the temperature, the stronger the current. The strength is measured by an electric meter.

This pyrometer has a thermocouple at the end of the two-foot wand, and a meter at the handle. The wand can be placed in a potful of molten lead (328°C.) and hotter materials, to an upper limit of about 1400°.

However, the melting point of iron is 1530°. And tungsten, used in making electric bulb wires, melts at 3400°. A wand-type pyrometer won't do for these high temperatures. We need a pyrometer that can measure temperature without touching the hot material. Such a pyrometer, called a *radiation pyrometer*, works by measuring the brightness of the light rays, or radiation, coming from glowing substances. It has a lens for focusing the radiation on one or more thermocouples. (Several together are called a *thermopile*.) The current is measured by a meter. Pyrometers, both the radiation type and the wand type, work well enough for measuring the temperature of

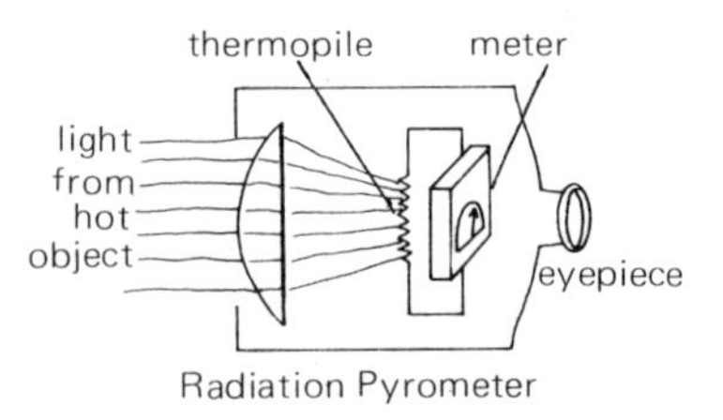

Radiation Pyrometer

nearby hot objects. For more distant work, such as measuring the temperatures of stars, *bolometers* are used ("bole" is Greek for beam).

Bolometers work on this principle: most metals, when heated, are poorer conductors of electricity than when cool. The principal part of a bolometer is a pair of blackened metal plates, usually of platinum. When a switch is turned on, electric current from a battery flows through the plates.

The bolometer is attached to a telescope where the eye lens is ordinarily located. The switch is turned on and the current through the plates is measured. Then the telescope is aimed at a star so that starlight strikes *one* of the plates. This plate becomes warmer, and therefore, conducts less electricity than the other. The hotter the star, the greater the difference in current between the two plates. This is measured by the electric meter.

There are many other instruments for measuring temperatures and for measuring the flow of heat. All of them work by the vibration of molecules.

Whole packages of molecules moving: mechanical energy.

Separate molecules moving helter-skelter in vibration: heat energy.

Molecules are made of *atoms*. The energy related to whole atoms is—

Chemical Energy

Here's a simple example of chemical energy. Leave a bit of egg yolk on a silver spoon for a half hour or so. A black deposit forms on the spoon. A chemical change has taken place. Atoms of shiny silver from the spoon have combined with atoms of yellow sulfur from the egg. A new substance has been formed: black silver sulfide.

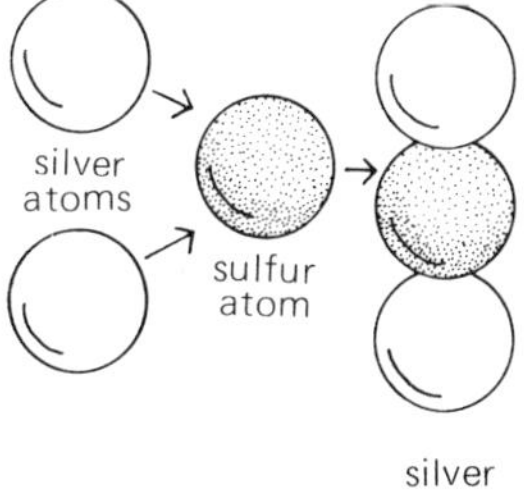

This is a diagram of what happened. Two

silver atoms became attached to one sulfur atom, forming a silver sulfide molecule. Trillions of these molecules make the black layer on the spoon.

Now, *how* did they become attached, and why did *two* silver atoms become attached to *one* sulfur atom?

Let's do a brief survey of atoms.

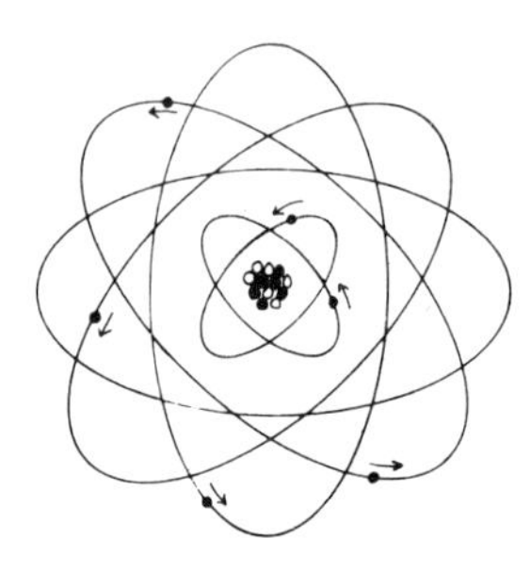

2 Electrons in First Shell
4 Electrons in Second Shell

1. All atoms consist of two groups of parts:
   a. The nucleus, made of a certain number of protons and neutrons.
   b. Electrons revolving around the nucleus. There are as many electrons as protons.
2. Electrons revolve in paths called *orbits.* A group of orbits is called a *shell.* Each shell is at its own special distance from the nucleus.
3. Each shell can contain a certain number of electrons and no more. The first shell, nearest to the nucleus, can hold no more than two electrons. The second shell can hold no more than eight. Some shells can hold more than eight. However—
4. The outermost shell can hold no more than eight electrons.
5. Some atoms have eight outer electrons. So their outer shell is filled. These atoms are called *stable,* or *inert,* which means inactive.

Sulfur Atom

2 Electrons in First Shell
8 Electrons in Second Shell
6 Electrons in Third Shell

Most atoms have fewer than eight outer electrons. Those that have seven, six, or five have an attraction that can pull other electrons to them. In that way they can complete their outer shells. For instance, a sulfur atom has six outer electrons. It could fill its outer shell by attracting two more, to make eight.

Now where could the sulfur atom get these two electrons? From two silver atoms. Each silver atom has only one electron in its outer shell. The pull of a silver nucleus on its single outer electron is weak. So the electron is pulled toward the sulfur atom's shell. Two silver atoms can each give one electron to the sulfur.

Silver atom

2 Electrons in First Shell
8 Electrons in Second Shell
18 Electrons in Third Shell
18 Electrons in Fourth Shell
1 Electron in Fifth Shell

When you know the number of electrons in the outer shell of an atom, you can predict how that atom will behave. Try it with this problem:

Sodium atoms have one outer electron.

Chlorine atoms have seven outer electrons.

Will sodium and chlorine combine? (Yes.)

How many sodium atoms will combine with one chlorine atom? One. (The combination is sodium chloride, which is ordinary table salt.)

Next problem:

Neon atoms have eight outer electrons.

Carbon atoms have four outer electrons. Will neon and carbon combine? (No, because neon's outer shell is filled. Therefore neon is inert. It doesn't give or take electrons; it doesn't combine with other atoms.)

Of course these ideas are quite rudimentary to chemists. They know more about atoms than just "The Rule of Eight." They understand a great deal about the behavior of matter: why gasoline burns, while water puts out fires; why stainless steel is stainless; how grape juice can be changed to wine—or vinegar; how to assemble atoms into a substance that is waterproof, not easily torn, cheaply and easily made into sheet form for raincoats, or twine form for ropes, or tube form for water pipes.

ENERGY IN CHEMICAL ENERGY. Every chemical change is also an energy change—energy is either taken in or given out.

Here are the two most important chemical changes on earth. In the first one, called oxidation, energy is given out:

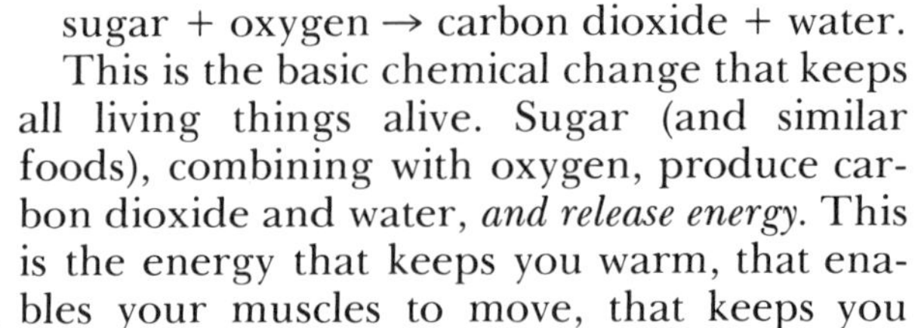

sugar + oxygen → carbon dioxide + water.

This is the basic chemical change that keeps all living things alive. Sugar (and similar foods), combining with oxygen, produce carbon dioxide and water, *and release energy.* This is the energy that keeps you warm, that enables your muscles to move, that keeps you alive in hundreds of ways.

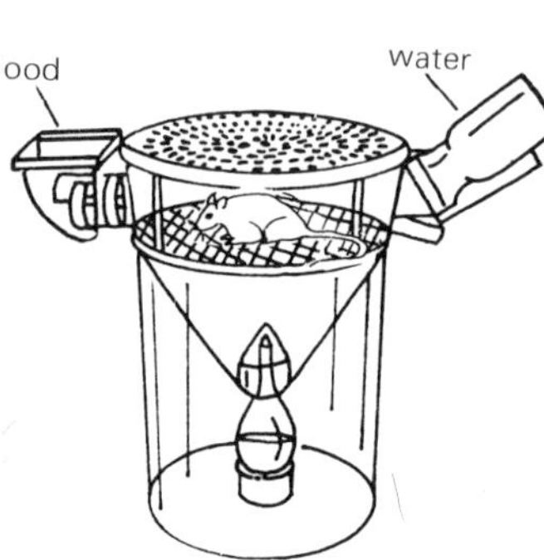

This mouse is getting a *basal metabolism* test. This is a measure of how efficiently he converts sugar and oxygen into carbon dioxide and water. Doctors do the same test on human patients.

In the second chemical change, equally important, energy is taken in.

carbon dioxide + water → sugar + oxygen.

This chemical change takes place in all green plants as they take in the energy of sunlight. The process is called *photosynthesis* (from the Greek for "light" and "put together"). The plants combine two "used-up" substances into sugar, and similar foods, such as starch. These can again be used by living things to keep them alive and going.

One of the important aims of plant breeders is to develop efficient food-producing plants. Such plants give increased amounts of food substance for the amount of land and fertilizer used.

Whole packages of moving molecules = mechanical energy.

Separate molecules moving helter-skelter in vibration = heat energy.

Atoms joining or separating = chemical energy.

Now we'll explore the parts of atoms called *electrons.* The energy related to electrons is—

Electrical Energy

The meter reader is finding how many kilowatt-hours have been used. Then the electric company sends a bill for something nobody has ever seen: electrical energy. What is electrical energy?

Think of a copper wire, made of copper atoms. Each atom of copper has twenty-nine electrons in the shells and twenty-nine protons in the nucleus. (There are also thirty-four or more neutrons, but neutrons are not involved with electrical energy.) A single proton is able to hold a single electron in place. Twenty-nine protons can hold onto twenty-nine electrons. Together, all the atoms have an equal number of protons and electrons. We say that atoms are electrically balanced, or neutral.

Suppose we force an extra electron into the first copper atom. Twenty-nine protons can't hold onto thirty electrons. What happens then? An outer electron in the first atom is pushed away toward another atom. Now *this* atom has one too many electrons, which causes one of its electrons to move toward still another atom . . . and so on, and so on. This stream of electrons, from atom to atom, is an electric current.

In the example above, only a few electrons are involved. Actually, in a wire there are billions of atoms crowded together and the electrons move in huge amounts. For instance, when you light a 100-watt bulb, about six *billion billion* electrons flow through the bulb every second! That's what you pay for to the electric company — to force electrons through the wires in your houses.

Some kinds of atoms allow their outer electrons to be pushed away easily. Usually these are atoms with fewer than four electrons in their outer shells. Such atoms are metals, and are good conductors of electricity. Copper, for example, is used for electric wires because it is a good conductor of electricity. Copper atoms have a single outer electron. So do silver and gold, but are, of course, too expensive to be used as conductors.

Scientists have thousands of different in-

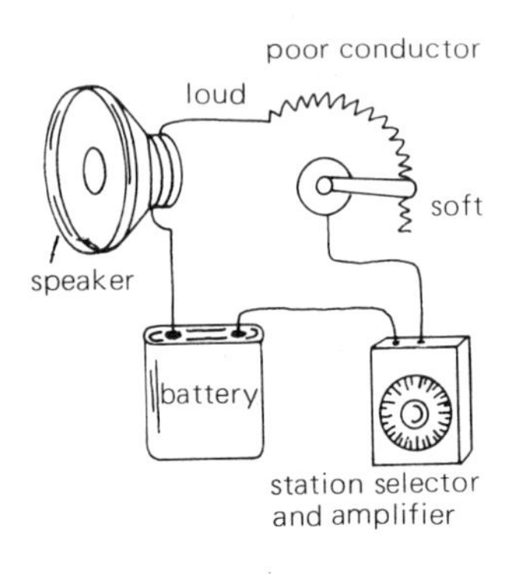

struments for controlling and measuring electric currents. Let's look at one controller and one measurer.

You have a controller in your radio, called the volume control. Its scientific name is *rheostat* (from the Greek for "current stopper"). Rheostats are used for many purposes besides making radios play softly or loudly. They control the brightness of light and the speed of electric motors. Here's the general idea of how they work.

An electric current from the radio battery flows through the rheostat on the way to the speaker. The current has to flow through the full length of a poor conductor. This weakens the current and makes a soft sound in the speaker. Turning the knob to the left allows the current to flow through a shorter length of poor conductor. The current is not weakened as much, so the sound is louder.

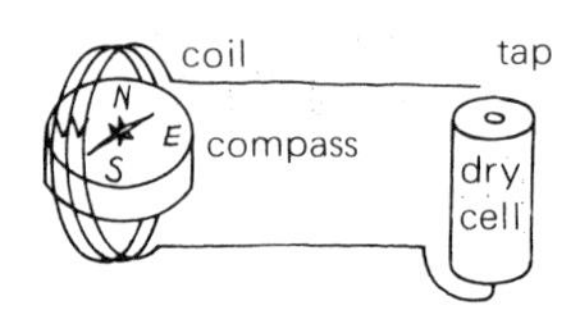

Current measurers, too, come in various sizes and shapes. According to their use they have various names such as voltmeter, ammeter, galvanometer, ohmmeter, wavemeter, decibelmeter and still others. But most of them have the same two basic parts: a coil of wire and a magnet. Here's a very simple one that you can assemble in a few minutes with a compass, a needle and a wire.

A compass needle is a magnet. The coil of wire is connected to the electric current to be measured. A weak current causes the magnet-needle to move slightly. A stronger current causes a greater movement.

In some electric meters, the magnet stands still and the wire moves. A pointer attached to the coil shows the measurement on a dial.

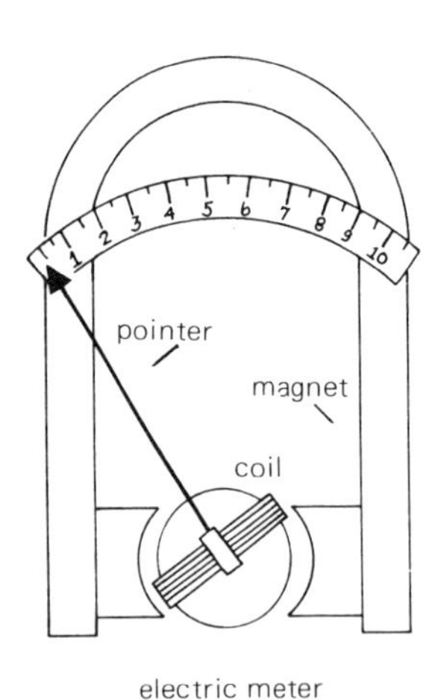

electric meter

Packages of molecules moving together = mechanical energy.

Separate molecules vibrating = heat energy.

Atoms joining or separating = chemical energy.

Electrons moving from atom to atom = electrical energy.

Up to this point, we have been exploring energy related to *things*: hammers, molecules, atoms, protons, electrons. Now let's take a deep breath and jump into a mystery.

Energy Related To Fields

A fully equipped parachute jumper is depending on three fields. He depends on the earth's *gravitational* field to pull him down to land. He depends on the earth's *magnetic* field to make his compass work. He depends on an *electromagnetic* field to carry the radio waves from his walkie-talkie to his partner's down below.

You have been living in a field all your life. The earth's gravitational field, or gravity, pulls you to the earth. It keeps the moon in its orbit around the earth. The moon's gravitational field pulled the Apollo astronauts to a moon landing when they approached near enough.

You have never seen a gravitational field, and neither has anyone else, because a field is not a thing. A field is a *condition* of the space between two things.

The space between a diving board and the water is in a certain condition. As a diver walks onto the board, step by step, the condition of that space changes. A force—some sort of pull or stretch—is set up between the diver and the water. The diver is willing to jump *up*, knowing he will not sail off forever. The condition of the space will bring him down.

This definition of a field—the condition of a space—may not seem very satisfactory, because it isn't. Scientists can't describe a field the way they or you can describe a pencil or a hamburger.

But scientists can do better in terms of *measurement.* They have measured the strength of the earth's gravitational force at sea level and at three inches above sea level. The difference is about one part in a billion. They have found that the earth's gravitational field is stronger above heavy rock layers, even when the layers are deep under the sea. The field is weaker above light materials such as subterranean beds of salt or sulfur. The differences are slight, but there are instruments called *gravimeters* that easily measure them.

You have a crude type of gravimeter in your home. It's called a spring-type bathroom scale. When you step on it, the earth's gravitation pulling you down causes a spring to stretch. The amount of stretch is shown on a dial. When your weight changes, the amount of stretch changes.

A scientist's gravimeter works in the same way, but the weight doesn't change. The weight is a metal cylinder inside the gravimeter. Gravity pulls down on it and stretches a spring. This is shown on a dial. When you take the gravimeter to a place where the gravity is lesser or greater, the spring is stretched less or more. Of course, such a gravimeter is much more sensitive and accurate than a bathroom scale.

Now let's explore another field that you have lived in all your life.

THE ELECTROSTATIC FIELD. When you take off a nylon shirt, it clings to you because of an electrostatic field. Shuffle your feet along a carpet and touch a metal doorknob or radiator: the shock you feel is a discharge in an electrostatic field. Another electrostatic discharge, on a much larger scale, is a bolt of lightning streaking between sky and earth. Do you own a shortwave radio receiver? Tune in

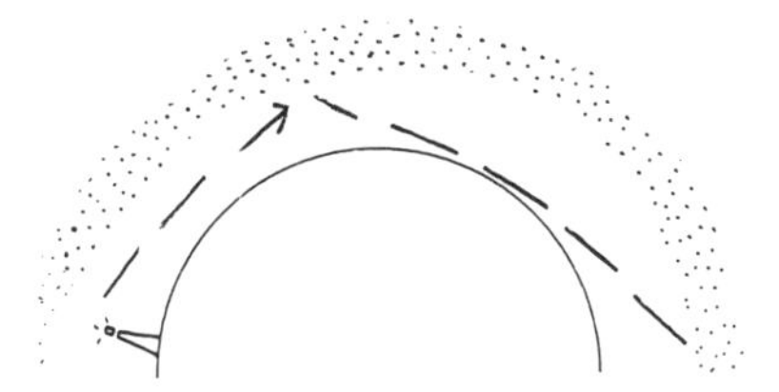

an Australian radio station, halfway around the earth. The radio waves, traveling up and away from Australia, are bounced toward you by an electrostatic field high up in the air.

All these fields are caused by crowds of spinning electrons or protons. Every electron and every proton has an electrostatic field around it. The field around an electron is said to be negatively charged, while a proton's field is said to be positively charged. You don't ordinarily notice these fields because every atom has an equal number of electrons and protons. Within the atom the fields cancel each other out. But if you can isolate some electrons or protons, you become aware of their electrostatic fields.

Let's try it.

Inflate a balloon and tie it at the neck. Balance a pencil across the edge of a book.

Rub the balloon across your hair a few times. This will temporarily rub off a few electrons (about a billion) from your hair onto the balloon.

Hold the balloon near one end of the pencil. Move the balloon a little, up and down, side to side. Watch the balloon pull the pencil, as if there were a gravitational field between them!

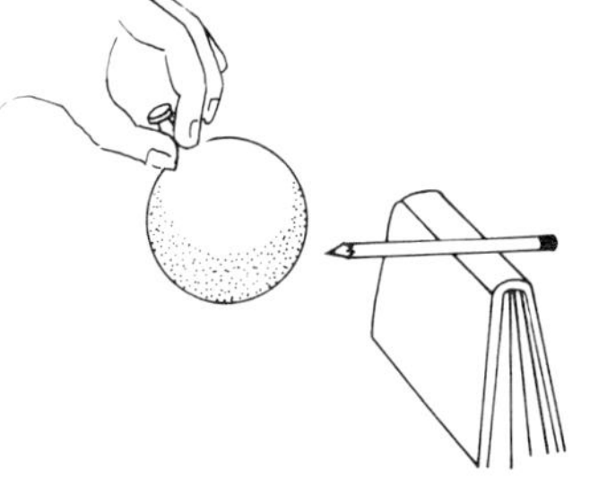

Well, isn't it a gravitational field? The answer is no, because you had to *rub* the balloon to produce the field. A gravitational field is there whether you want it or not.

There's another reason as well. Try this:

Inflate another balloon and tie it at the

neck. Rub both balloons on your hair. Lay one balloon on the table and hold the other near it. They *repel* each other! The electrostatic field of one balloon is repelling the electrostatic field of the other. But gravity fields never repel each other.

In these experiments, you transferred electrons onto the balloons. These electrons came from your hair, leaving protons without partners. After a while, the electrons drifted from the balloons into the air and back to your hair, attracted by the protons. The electrons settled back into place in the outer shells of atoms. There they took on another motion: revolution. Each electron whirled around a nucleus, millions of revolutions per second. This whirling motion produced another kind of field. . .

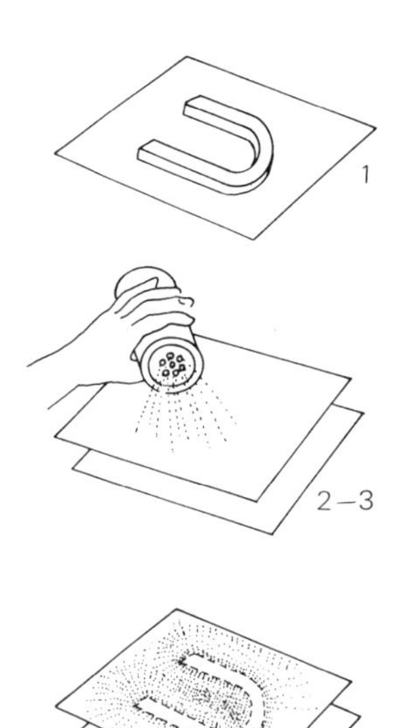

THE MAGNETIC FIELD. Every electron revolving around a nucleus creates a magnetic field. The magnetism of a magnet comes from the billions of little magnetic fields produced by billions of whirling electrons. Added together, these tiny fields make a big field. The shape of this field can be shown by sprinkling iron filings on a piece of cardboard placed over a magnet.

Well, your pencil contains lots of electrons—why isn't your pencil a magnet? Or this book? Why aren't *you* a magnet?

Lots of little magnetic fields can add up to no magnetic fields. An electron traveling clockwise around a nucleus cancels the magnetic field of an electron traveling counterclockwise. And that's what happens in most atoms in most substances: canceled fields. But if an atom does have some electrons with uncanceled fields, these fields do show up. Ordinary magnets, made of iron or steel, have atoms with uncanceled fields.

However, it isn't enough to have *atoms* with

uncanceled fields. The atoms in turn have to be lined up so that their *fields* don't cancel other fields. That's what you do when you stroke an iron nail or steel needle with a magnet. You line up the uncanceled fields so that they add up.

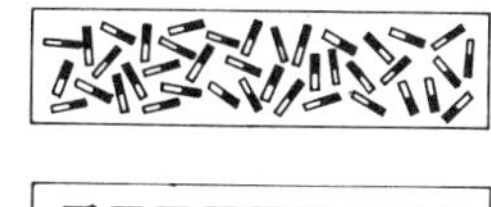

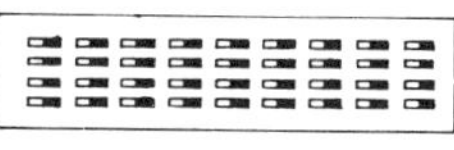

Try it yourself. Take two sewing needles and test them for magnetism. Can you pick up one needle with the other? Then place them side by side, point to point and eye to eye. Stroke them with a magnet, with a combing motion, about fifty times. This lines up the uncanceled fields. Now you can pick up one needle with the other. What happens when you touch the point of one needle with the eye of the other? How about point to point? Eye to eye?

Hold the needles with pliers and heat them red hot. The heat causes the atoms to vibrate, to shake in all directions. When the atoms cool, their magnetic fields are no longer lined up neatly. Can you use the needles as magnets?

THE MAGNETIC COMPASS. Scientists have many uses for magnets. Perhaps the oldest use is in magnetic compasses, to point to magnetic north and south. A magnetic compass has a magnetized needle that is lined up by the earth's magnetic field. Scientists are not sure of the cause of the earth's field. They believe that the slow circulation of molten metal, deep down in the earth's core, has something to do with it.

The earth's magnetic field doesn't have a neat, regular shape. It twists toward a deposit of iron ore, even if the ore is deep in the ground. Other minerals also have their effects. The bedrock under oil fields affects the field, and oil prospectors make use of this. An instrument called a *magnetometer* is hung under an airplane. It measures the strength and direction of the magnetic field as the airplane flies back and forth in a regular pattern marked on a map. Oil geologists, studying such a *geomagnetic map,* can spot likely looking areas for test drilling.

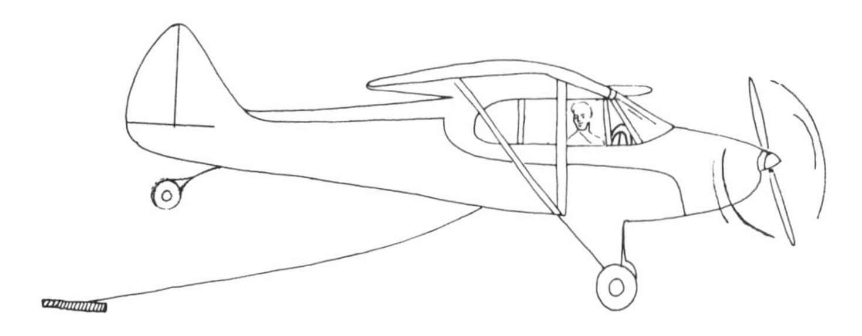

ON-AND-OFF MAGNETS. The magnets in compasses and magnetometers are called *permanent magnets* because their fields are lined up permanently. Unless you destroy their alignment (as you did by heating the needles) they will continue to have a magnetic field around them.

Sometimes we want a magnet whose field can be turned on and off—a temporary magnet. A door chime has such a magnet. So has the huge crane that can pick up a ton of scrap iron, swing it over and drop it into a freight car. So do many machines for handling dangerous materials in scientific laboratories.

The fields of temporary magnets are produced by an electric current. Turn a switch and the current flows. Whenever an electric current flows, a magnetic field builds up around it. Stop the electric current and the field collapses. You can see this for yourself by doing the experiment on page 95.

The magnetic field that surrounds the wire is rather feeble. But if we coil up a long wire into a shape like this—

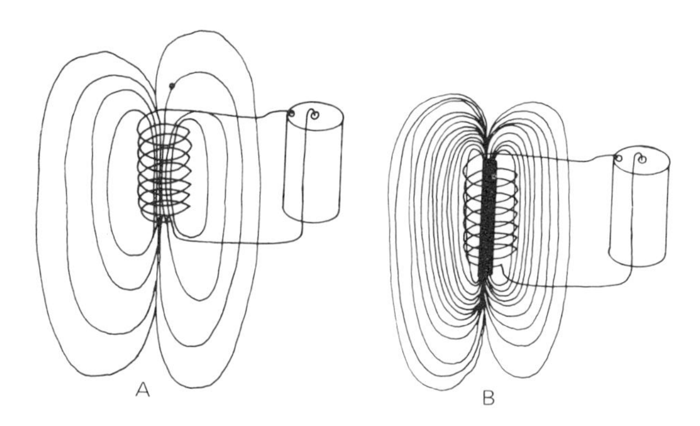

we crowd the field into a smaller space, and get a more concentrated field. We can concentrate it still further by attracting the field with an iron bar, called a core, placed inside the coil. The current flowing through the coil, surrounding the core, makes up the very useful device called an electromagnet.

Electromagnets are the main working parts of electric motors. They turn the hands of your electric clock. They work the freezing machinery in your refrigerator. They turn the turntable on your record player. Electric motors run hundreds of devices in scientific laboratories. A tiny motor moves the paper chart on which a pen traces a graph of the electrical energy of brain waves. A bigger motor keeps the Mount Palomar telescope pointing to a selected star while the earth rotates.

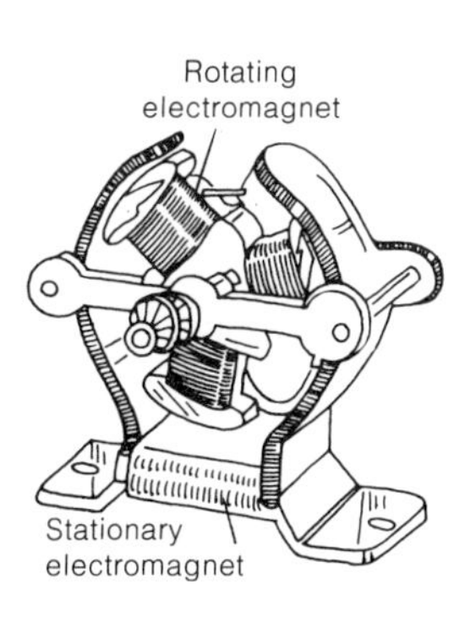

ELECTROMAGNETIC FIELDS IN SPACE. Radio transmitters work by sending out waves through an electromagnetic field. These are called electromagnetic waves, or Hertzian Waves. You can't see or feel them, but you can make them. You will need a transistor radio, a battery and a piece of wire.

First make a few waves yourself, as this

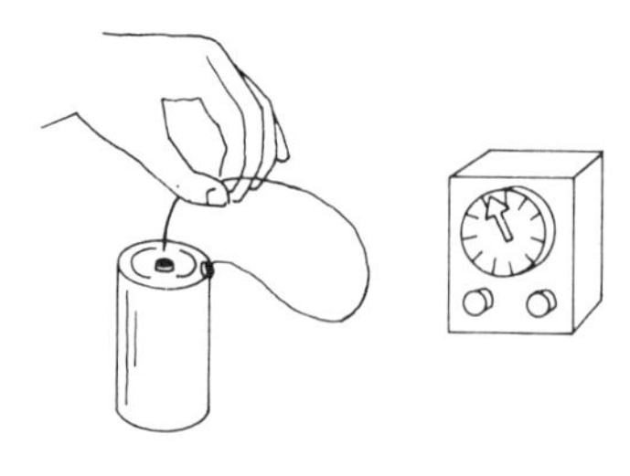

diagram shows. Turn on a transistor radio and find a quiet place on the dial. Tap the wire at the battery on and off. Listen to the clicks coming out of the radio!

The sounds were made by radio waves spreading out from the battery wire. Some of them struck a coil of wire in the transistor radio and were changed into clicking sounds. In the same way, radio waves from a walkie-talkie spread out. Some of them bump into the other walkie-talkie.

What about the waves that spread out in other directions? They keep traveling out into space. A second and a half after you tap the wire some of the waves will reach the moon! Eight minutes later some of them will bump into the sun, while the others will keep going—five hours across the solar system, four years to the nearest star—and still on and on. What a breathtaking thought.

Now let's take a brief look at radio in a broadcasting station, in your transistor radio, and in a scientific laboratory.

Let's start by looking in at a radio station, for example, New York City's municipal radio station WNYC, which transmits at 830 *kilocycles* (kilohertz), which means 830 *thousand* cycles, or waves, per second. At the start of the broadcasting day, an engineer turns on the transmitter. Immediately a stream of electrons travels up the WNYC antenna and back, 830,000 times per second. Each round trip ("cycle") sends one electromagnetic wave out into space. A diagram would look like this.

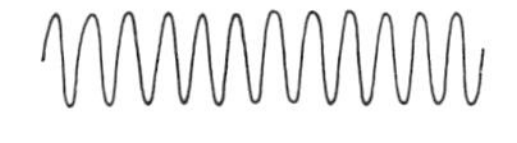

If you turn on your radio at this time, you'll hear nothing, because you can't hear waves of such a high *frequency*. They are called *carrier waves*, because they will carry other waves as soon as sounds are made at the microphone, in the broadcasting station.

Then the announcer says something, such

as "Good morning." His sound waves could be diagramed like this:

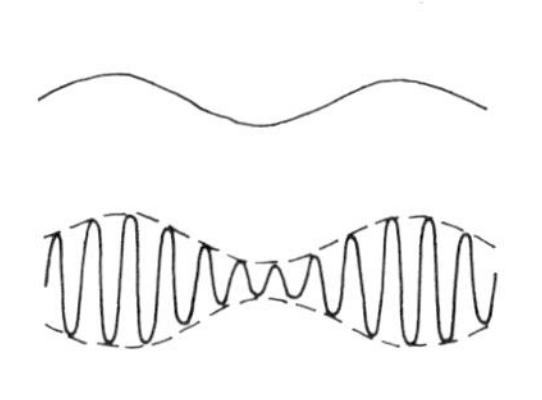

These waves have a lower frequency, which can be heard by human ears.

The radio transmitter combines the two sets of waves into a new set. This new set contains the "830-kilocycleness" *and* the "voice-vibrationness."

The combined waves are sent up the radio station antenna, through space in all directions. Some of them strike your radio receiver. If you set your dial at 83 (which stands for 830,000) the waves are allowed in. Otherwise they are kept out. Let's say the dial is set at 83. Then the waves are changed so that they can make sounds come out of an earphone.

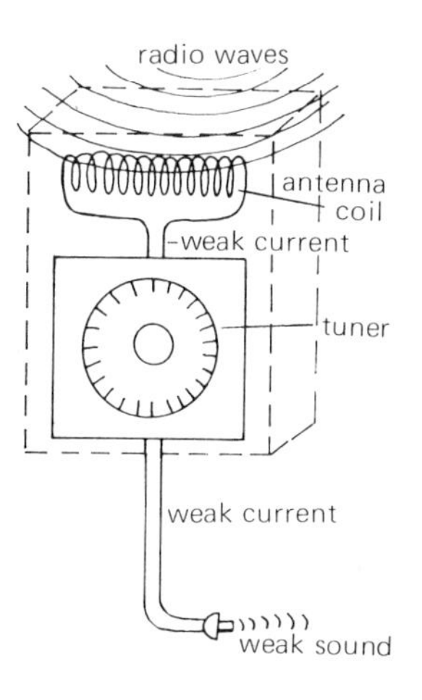

Maybe you've listened to an inexpensive little earphone radio receiver called a *crystal set.* Such a receiver works without batteries or house current. The electric current for the earphone is made by the radio waves themselves. As they sweep across a coil of wire called the *antenna coil,* they set up a feeble electric current. This is just enough to make a weak sound in the earphone. To get a louder sound, you use a radio that contains an—

AMPLIFIER. The job of an amplifier is to take in a feeble current and send out a strong current of the same frequency. The early amplifiers had parts, called *tubes,* that looked like small electric bulbs. Most modern amplifiers have, instead, one or more *transistors.*

A transistor is a little metal "sandwich." The "bread" is two metallic crystal plates. The "filling" is a plate made of a different metallic crystal.

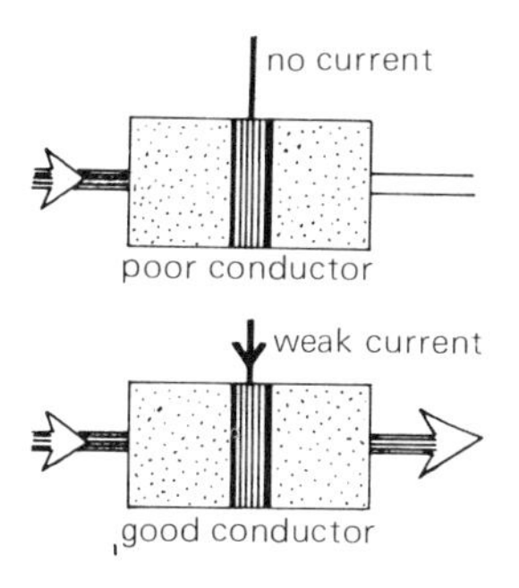

A transistor has an unusual property. When you try to pass an electric current *through* it, from one outer plate to another, it's a poor conductor. But when you connect a weak electric current to the edge of the inner plate, the whole transistor becomes a very good conductor. Stop the weak current, and it becomes a poor conductor again. Let's see how this property is used in a one-transistor amplifier. (Most amplifiers have more than one transistor.)

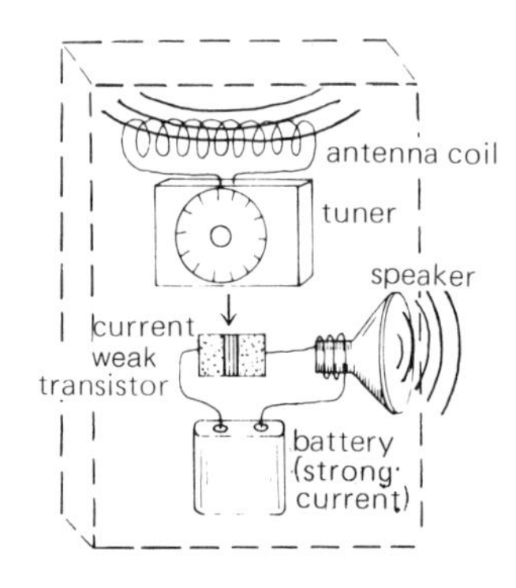

Here's the transistor, connected to a battery and a loudspeaker. If the dial is not set to an operating radio station frequency, there will be no weak current from the antenna coil to the inner plate of the transistor. Therefore, the transistor would be a poor conductor. No current would flow from the battery to the speaker.

Next, turn the dial to the frequency of a station that's "on the air." A pulse of weak current flows from the antenna coil to the inner plate. The transistor becomes a good conductor, and a strong current flows from the battery to the speaker. This makes a single sound vibration. A weak current (from the antenna coil) can control a strong current (from the battery). One after another, the pulses of weak current produce strong sound vibrations.

This system for using a weak source of energy to control a strong source, has many uses for scientists. Maybe you've had a heart examination with an electrocardiograph ma-

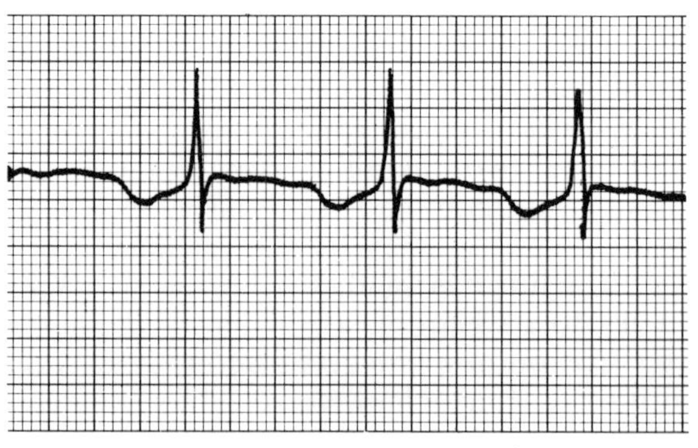

chine. Your heartbeat sends out pulsing electric currents that spread throughout your body and reach your skin. These signals are transmitted through wires to the machine. There they are amplified from about .003 volts to three volts, an amplification of 1,000 times. The amplified current is strong enough to operate an electromagnet that moves a pen. The pen makes a record of your heartbeat on a moving chart.

Another important scientific use for amplifiers is in a process called *telemetry,* which comes from the Greek meaning "far-measuring." A Venus probe lands on the surface of that planet and begins to take all kinds of measurements: temperature, pressure, humidity, light intensity, and many others. These measurements are carried by radio waves to an antenna on earth. The radio waves set up little feeble currents in the antenna at a few *millionths* of a volt. These are amplified several million times so that they can operate a machine that records the measurements.

But how are the Venus measurements taken? The temperatures are taken by an instrument that measures and transmits temperature readings. The "thermometer" is a rod called a *thermistor.* A thermistor conducts electricity according to temperature. The hotter it becomes, the better it conducts electricity, but it becomes a poorer conductor as the temperature falls. The thermistor is connected to a battery and a special transmitter. A low temperature (weak current) causes the transmitter to send few waves per second, like this:

A high temperature (strong current) causes the transmitter to send many waves, like this:

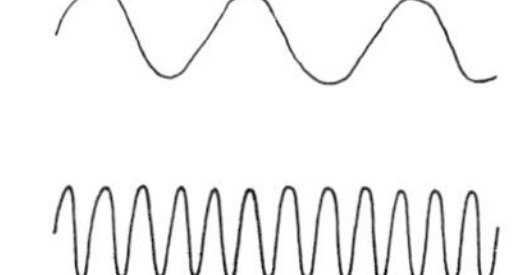

The waves are received on earth and amplified. A computer converts "waves per second" into temperature readings.

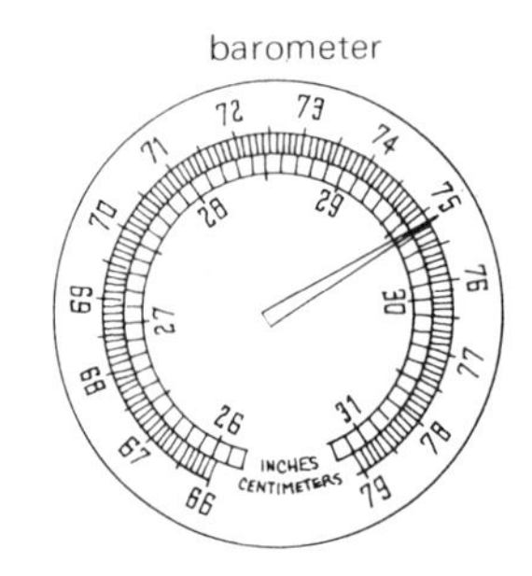

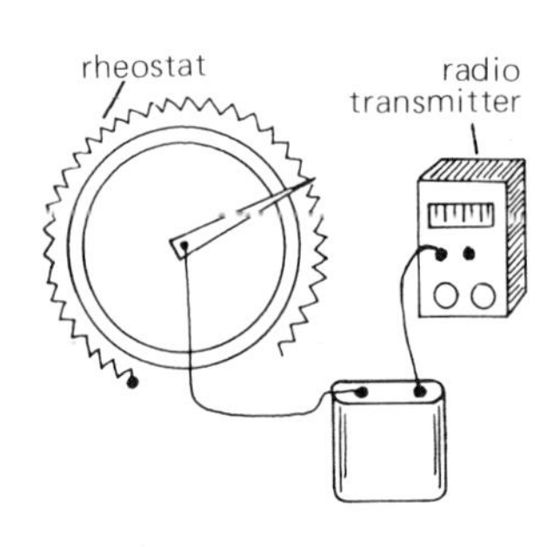

Here's an ordinary *barometer*. A pointer shows the air pressure on a dial. Suppose, instead of a dial, you had a curved section of a rheostat. You could connect it to a radio transmitter to telemeter information on pressure back to earth. Can you finish this hookup to make such a telemetric system?

HIGHER FREQUENCY WAVES. Radio waves are sweeping through you right now, but you can't see or feel them. The same is true for the next higher frequency—television and radar waves. But you would certainly feel the next higher frequency, *microwaves*, if you put your hand in a microwave oven.

A microwave oven cooks and bakes much faster than an ordinary gas or electric oven. The main part is a "wave generator" that makes the high frequency waves called microwaves. These can pass right into the raw food. Electromagnetic energy of the waves causes the food molecules to vibrate faster—that is, to heat up. So the whole piece of raw food—a chicken, a fish, a piece of beef—is heated evenly, inside and outside at the same time.

The even heating and speed make microwave ovens useful in scientific laboratories.

The next higher frequency waves are the *infrared* waves. The warm feeling of sunlight comes mostly from infrared waves. You too are making them, right now, and you can feel them. Curve your palm and hold it near your forehead, without touching. Feel the warmth made by infrared waves traveling back and forth between the molecules of your palm and forehead.

These are genuine electromagnetic waves. As your warm skin molecules vibrate, so do their electrons. Whenever electrons vibrate they send out electromagnetic waves, one wave per vibration.

When your hands get cooler (fewer vibrations) the frequency is lower. Colder and colder, you still keep on making waves, of lower and lower frequency. And so does everything else—plants, omelets, oceans—even icebergs!

There is a bottom limit to this cooling. It's called absolute zero, the temperature at which molecules have entirely stopped vibrating. At this bottom limit, no infrared waves at all are made. Absolute zero is 458.9°F. below zero or 273°C. below zero.

There is a whole branch of science called *cryogenics* (from the Greek for "cold maker") that deals with the behavior of materials in extreme cold. The first piece of equipment you would need to experiment in cryogenics is a thermometer. But liquid thermometers freeze long before absolute zero is reached, and bimetallic strips become stiff and fragile. However, there are other ways of measuring low temperatures. Bolometers, described on page 85, can measure infrared energy of low frequency. Thermistors, described on page 101, can also be used.

The name infrared means "below the red." If a substance such as a bar of iron is heated to higher and higher temperatures, it gives off infrared waves of higher and higher frequencies. Finally it reaches the temperature where it begins to glow with a dull red light. At that point it has gone past the infrared frequencies, to the frequencies that are called—

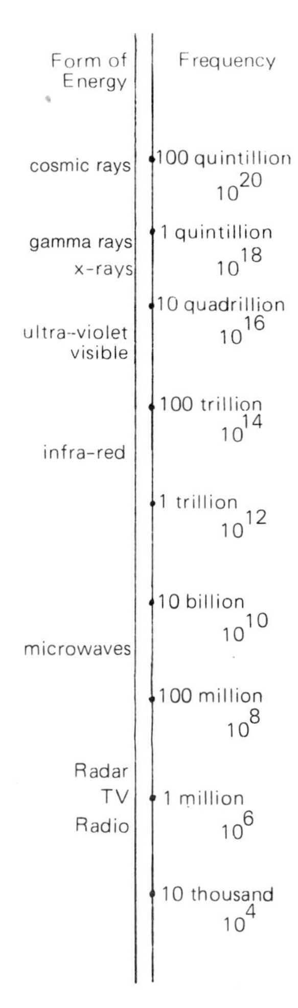

VISIBLE LIGHT. The energy of visible light shines out of stars, the sun, electric lights, campfires, fireflies, candles. Visible light consists of electromagnetic waves. These are the same *kind* of waves as radio, TV, microwaves and infrared waves. But they are produced in a different part of the atom—in one or more of the interior shells.

Electrons usually revolve in their orbits at a

certain distance from the nucleus. The electrons are said to be in the *ground* state. But if an atom is heated hot enough, its vibration will fling some electrons away a little distance. These electrons are in an *excited* state. Immediately they are pulled back by the attraction of protons in the nucleus. In a light source such as a candle or electric bulb, this happens again and again, trillions of times per second. Electrons are forced into an excited state and pulled back by the attraction of the nucleus.

It takes energy to force an electron out to a higher orbit, against the powerful pull of protons in the nucleus. And each time the electron snaps back to a lower orbit, it gives back the same *amount* of energy, but in a different *form*. The energy is returned in the form of a flash of light, or *photon*. Trillions of photons per second add up to the yellow glow of a candle flame, the many-colored flickers of a burning driftwood log, the reddish light of a neon sign, the blues and violets of late sunsets, the white light of the Dog Star, Sirius.

When photons are emitted, they are accompanied by electromagnetic waves that travel out into space. The *color* of light depends on the *frequency* of these waves. The lowest frequency (about 420 trillion waves per second) gives red light. The highest frequency, about 780 trillion, gives blue light. The whole range, or *spectrum*, of visible light goes from red to orange to yellow to green to blue to indigo to violet.

Trillions of waves per second

750 600 500 430 400

violet . . . . .indigo . . . . . green . . . . . yellow . . . . . orange . . . . .red . . . . .infra-red

The color of a substance is the color of the light it reflects. A lemon is yellow because it reflects yellow light—light whose frequency is about 550 trillion waves per second.

When we speak of the colors of visible light, we mean "visible to people." Most animals can't see colors at all—everything looks black or white or shades of gray. Some animals can see colors that are invisible to people. Bees, for example, can see light of a higher frequency than our visible spectrum, beyond violet, to *ultra*violet. We can't even guess what ultraviolet looks like to bees. Neither can bees guess (if bees could guess) what deep red looks like to us.

We use light energy as a way of making things visible. But light can also be used as a tool, as a scientific instrument. Imagine being able to drill a hole in a razor blade twenty feet away by just shining a narrow beam of light at it. Sounds like science fiction, but scientists can do it easily with an instrument called a—

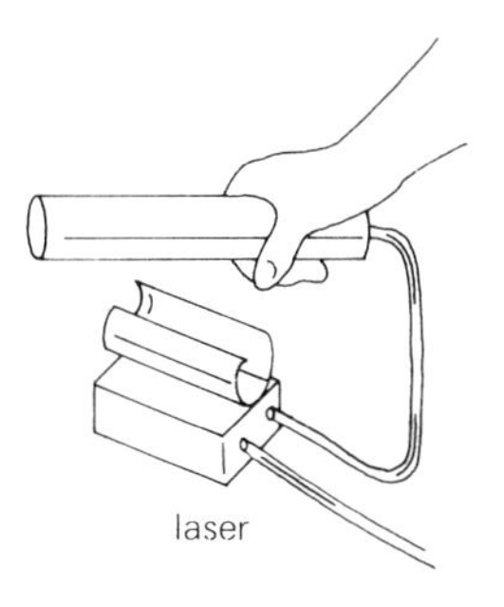

LASER. This word comes from *L*ight *A*mplification by *S*timulated *E*mission of *R*adiation. This isn't a very helpful piece of information as yet. But imagine this situation: We want to break down the heavy wooden gate of a castle in which a fair maiden is imprisoned. We can order each of our dozen foot-soldiers to hit the gate with sticks, as often and with as much power as his strength allows. Or we can get one large stick for them all, a battering ram, so that all their strengths are combined and they all strike the gate at once.

Ordinary light from an electric bulb is a mixture of many frequencies of light. Shine such a light at a wall and you're hitting the wall with a whole mixture of frequencies, like the soldiers with their sticks, each hitting at his own rate. But a laser produces light waves of exactly the same frequency, exactly in step, so their energy is delivered together, like a bat-

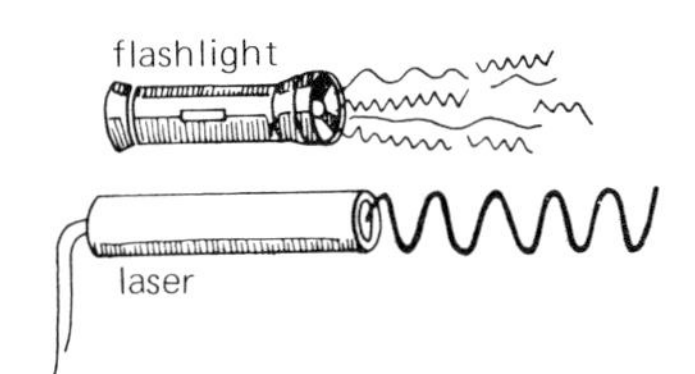

tering ram. A lot more strength—enough to punch a hole in a razor blade—and more.

A laser beam is very powerful, and yet it's controllable. You know how the ordinary light from a flashlight or an auto headlight spreads and scatters, even though we try to control it with curved mirror reflectors. But a single frequency can be controlled—and that's what comes out of a laser light. The beam, no thicker than a pencil, shines out straight and narrow, without spreading, for many miles. So surveyors can use a low power laser beam as a kind of "tape measure" to measure long distances along a straight line. The length of the line is measured by an instrument that counts the number of waves in the line.

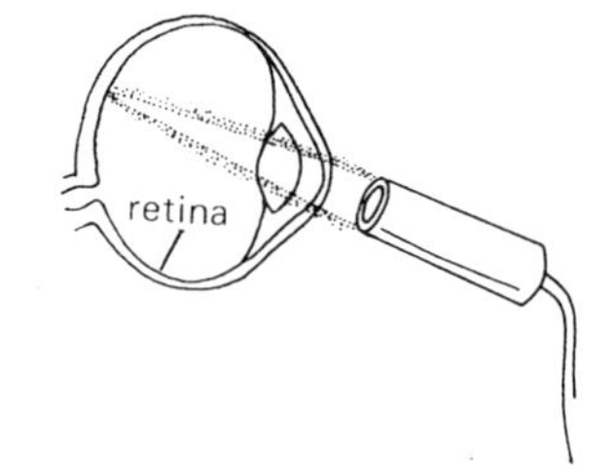

Power and control make the laser useful in many ways for scientists. An eye surgeon, for example, can form a beam of laser light as thin as a pin point. He can aim it right at a place where a part of the rear layer of the eye, the retina, has become detached from the eyeball. Several quick flashes of the powerful pinpointed beam, and the retina has become "welded" in place again.

Lasers are being developed for all kinds of uses: in surgery for destroying diseased tissue; in industry for welding tiny electronic parts; in communication for acting as carrier waves; and for many other purposes.

X RAYS. There is a form of electromagnetic energy with a frequency still higher than visible and ultraviolet light. It's the very, very high frequency invisible light called X rays.

Like other frequencies of light, X rays are given off when electrons snap back to their customary orbits after being forced out in an excited state. But these electrons are only in the innermost shells, closest to the nucleus.

Because they are so close, it takes an enormous amount of energy to shove them out. Therefore, when they snap back, they give off waves and photons with an enormous amount of energy in them—X rays. With so much energy, X rays can force their way through many substances—skin and muscle, for example—but they have a more difficult time getting through denser material such as bone. It's this difference that is recorded on an X ray picture.

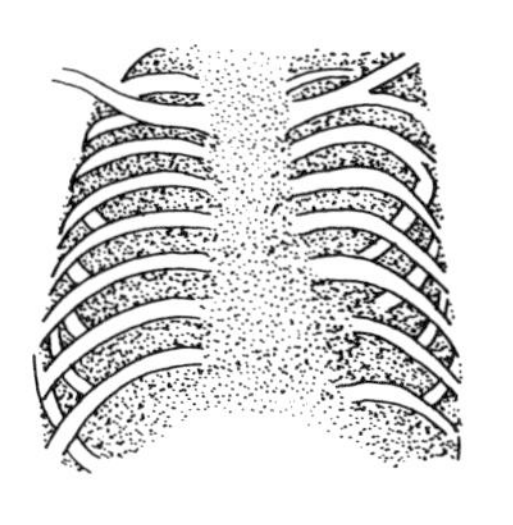

Such differences are also found in other kinds of electromagnetic waves. For instance, radio waves can pass through brick and concrete but visible light waves cannot. That's why you can play a radio inside a brick building, but you can't see through the walls. Think of other materials in relation to the different wave groups. Are any waves stopped by glass? By wood? Do radio waves pass through *all* materials? (Turn a transistor radio on to some music, then put it into an empty metal coffee can or wrap it in aluminum foil. Then try a paper bag.)

ELECTRONS MAKE WAVES. We have been looking at electromagnetic waves—radio and television, infrared, microwave, radar, visible light, ultraviolet, X ray. These waves are different in one way: their frequency. They are all alike in two ways: (1) they are forms of *electromagnetic energy*; (2) they are caused by the movement of electrons.

Now let's take another look at electrons. Maybe you want to clear up some questions such as these:

What makes electrons spin? Answer unknown to date. Scientists say "It's a fundamental property of matter." This is almost like saying "They spin because that's what electrons do."

But let's be patient. Scientists have discovered more about energy and matter in the last hundred years than in the previous hundred thousand years.

What makes electrons revolve around the nucleus? Same answer as for the first question, and same request for patience.

What keeps electrons from bumping into each other? Electrons repel other electrons. Their electrostatic fields keep them apart.

What keeps electrons from being pulled into the nucleus? The whirling, revolving motion of the electrons keeps them away.

Now let's examine waves again. Notice the change in the emitted waves as we come closer and closer to the nucleus:

The wave frequency becomes higher. The lowest frequency waves are radio waves; they come from the outermost electrons. The highest frequency waves are X rays. They come from the innermost electrons.

The energy of the waves becomes greater. The weakest waves are radio waves. The most energetic waves are X rays.

Suppose we move still closer, *into* the nucleus. Will we find *still higher* frequencies, *still greater* energy?

Yes. The nucleus produces waves of enormously high frequency, called *gamma rays.* The nucleus produces the most powerful form of energy: nuclear energy.

What happens in nuclei when they give out waves and energy? Nuclei are so small that it's very hard to investigate what goes on inside. Imagine an atom of carbon enlarged to the size of a football field. The nucleus of the carbon atom would be the size of a cherry pit, at the center. The outer electrons would be the size of apple seeds, at the outer rim of the field. Even much larger nuclei, such as those of uranium and plutonium, are too small to be seen with the highest power microscopes.

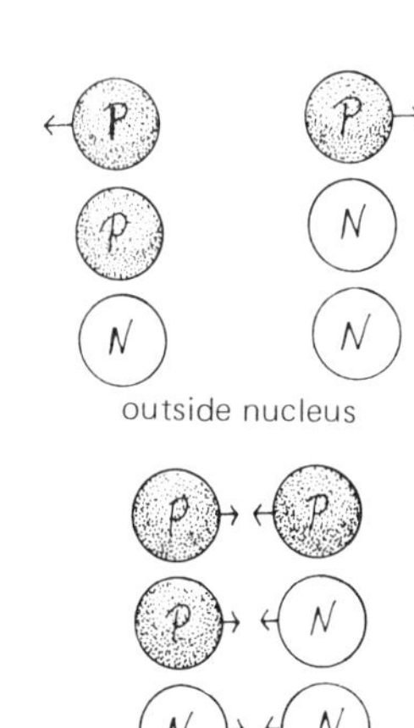

Nuclei contain mainly protons and neutrons. If you experimented with protons and neutrons *outside* the nucleus, you would get these results:

Proton to Proton: repulsion
Proton to Neutron: no effect
Neutron to Neutron: no effect

But look at what happens *inside* the nucleus:

Proton to Proton: attraction!
Proton to Neutron: attraction!
Neutron to Neutron: attraction!

Still more puzzling, these attractions occur only at a certain short distance. Bring the particles still closer and they repel! Pull them farther apart and they behave as they would outside the nucleus. Why? To date—*answer unknown.*

THE "BLACK BOX." Right now all over the world, scientists are studying nuclei for two main reasons: (1) To a scientist, the label *"answer unknown"* is enough of a challenge. (2) To all of us, the answers could bring some extraordinary benefits:

Unlimited safe energy at very low cost.

No more air pollution from burning fuels.

No more destruction of the earth by mining or drilling.

No more worries about using up the earth's reserves of coal and oil.

How can we get all these benefits out of nuclei? The details are much too complicated to be explained briefly. But we can do a "black box" examination of the problem.

"Black Box" is a scientist's name for a device that can be described like this:

We know what goes in.

We know what comes out.

But we don't know what goes on inside.

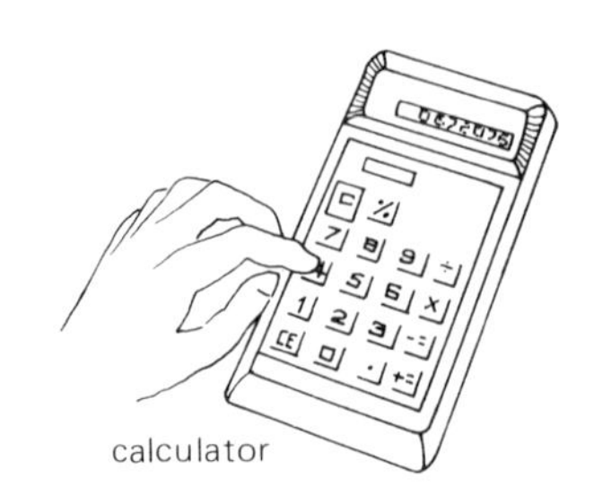
calculator

For instance, an electronic calculator is probably a black box for you. You know what goes in: electrical energy from a battery and mechanical energy from your finger. You know what comes out: some lighted numbers that give the answer to a problem. But you don't know what happens inside.

Of course, a black box for you is not a black box for the electronic engineer who designed the calculator. Neither is an auto engine a black box to an auto mechanic, although it is to many drivers. (Gasoline and air go in—mechanical energy, heat energy, and exhaust gases come out. But what happens inside?)

Nuclear Energy

The nucleus of the atom is almost a black box to scientists. They have many unanswered questions about nuclei. Their most important black box problem is how to achieve this:

4 protons → black box → 1 helium nucleus + ENERGY

Suppose you could put four protons into the right kind of box. There would come out one helium nucleus and an enormous amount of energy. "Enormous" means, for example, that the protons in a cupful of water could provide your car with as much energy as 2,000 gallons of gasoline! That's enough for six round trips across the United States.

It's easy enough to collect protons and bring them *near* each other. But when we bring them still closer, the protons repel each other with tremendous force.

Is there a way of overcoming the repulsion?

Yes. Fire the protons at each other at great speed.

Is there a way of achieving a great speed?

Yes. Heat the protons to a very high temperature, to about 50 million degrees centigrade. The protons' furious vibration will slam them into each other.

Is there a way of achieving such a high temperature?

Yes. The explosion of a nuclear bomb produces an even higher temperature. But obviously a bomb has no place in our program for safe, low cost energy.

Scientists are experimenting with beams of laser light (see page 105). When several powerful laser beams are pinpointed at one tiny cluster of protons, their temperature rises sharply. The protons slam into each other with enough force to join and stay together. This joining is called *fusion* (from the Latin word for "melt").

To achieve useful fusion, the process must be done on a large scale, not with a tiny cluster of protons. And there we meet the big, unsolved black box problem: how do we build the furnace?

No materials can stay solid, or even liquid, at fifty million degrees. Perhaps there is a way of setting up a repulsion field in the furnace walls. This field would keep the hot stuff suspended in space, away from the furnace walls.

In the meantime, until fusion is achieved, we have another way of getting energy from the nucleus: by splitting heavy nuclei. This is called nuclear *fission* (from the Latin word for "split"). There are more than a hundred electric powerhouses that work by the fission of uranium or plutonium.

Then why do we bother looking for fusion, if we already have fission at work? Because fission has many serious problems. The most important are these:

THE RUNAWAY DANGER. There are control devices in a fission electric generating plant. These devices are like the gas pedal of a car, to regulate the speed of the machinery. If the gas pedal gets stuck while the car is at high speed—danger! The fission control devices in an energy plant have all kinds of safeguards, but suppose the safeguards fail, one after another? Then a runaway situation could develop. The overproduction of energy would melt the fuel. Explosions would shower the land and air with deadly fission products.

THE WASTE PRODUCT DANGER. When a campfire burns down, it leaves a harmless waste product—ash. When nuclear fuels undergo fission, they leave several highly dangerous radioactive waste products. These give off high-energy rays that can damage or kill living things. They continue to do so for over a hundred thousand years!

How do we store them until they are no longer dangerous? There are several methods in use now, that are called "almost absolutely certainly" safe. Is that safe enough for you, or for someone who will be born fifty thousand years from now?

Clearly, nobody is altogether happy about nuclear energy from fission. The runaway and waste product dangers are always present.

Neither of these dangers exists for fusion. A control failure would simply stop the fusion; and there are no harmful waste products to store. So scientists have good reason to keep trying to achieve fusion in a practical way.

And they have good reason to be hopeful, because nuclear fusion already exists, on a very large scale! The enormous energy of sunlight and starlight seems to come from nuclear fusion. It's a complicated process, but here is a part of the black box description:

4 protons → black box → 1 helium atom + ENERGY

Just what we were looking for!

Energy Everytime, Everywhere

All the happenings of your day are involved with energy. When you wind your alarm clock, you put mechanical energy into it, in the potential state. The energy comes out a bit at a time—you can hear it ticking. If it's an electric clock, electrical energy is converted into mechanical energy. At breakfast you take in some potential chemical energy (eggs and toast). During the day you gradually convert it into mechanical energy (muscle motion) and heat energy.

It's interesting to look at the world as a series of energy changes.

Try it for a while!

	Mechanical	Heat	Chemical	Electrical	Electro-magnetic	Nuclear
Mechanical	*Hammer*					
Heat						
Chemical		*Candle*			*Candle*	
Electrical	*Electric drill*					
Electro-magnetic					*Telescope*	
Nuclear						

You could begin by finding a few fillers for the blank spaces in these columns. Some have already been filled in, as starters. For instance, an electric drill converts electrical energy into mechanical energy. A candle converts chemical energy into one form of electromagnetic energy (visible light) and heat energy. A hammer converts the mechanical energy from your arm muscles (at the handle) into mechanical energy that drives in the nail. A telescope focuses light energy from a distant star into light energy in your eye.

Here are a few energy converters, waiting to be placed in boxes: microscope, electric bulb, screwdriver, doorbell, guitar, match, centrifuge, flashlight, telephone, handsaw, jet engine, light meter, nuclear generating plant, record player, stove, toaster.

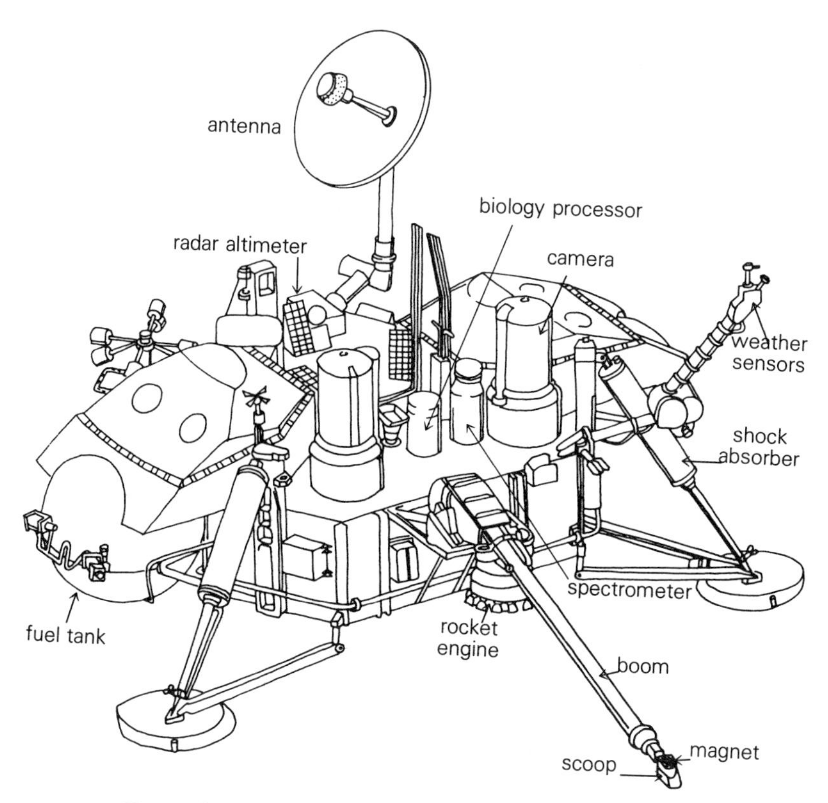

Matter, Time, Space and Energy on Mars

Meet Viking lander #1. This extraordinary package of scientific instruments was launched inside a spacecraft on August 20, 1975. Its twin, Viking lander #2, was launched three weeks later. Both were sent on an eleven-month voyage to Mars, to make landings at two selected places on the planet. Both were designed to seek answers to many questions, adding up to these three broad questions:

1. What are the conditions of the surface and atmosphere of Mars?
2. Are such conditions likely to support life on Mars?
3. Is there life on Mars?

You might expect the scientific instruments to be extremely complicated—and indeed they are. There are more than a half-million transistors, as well as many other parts—rocket engines, electric generators, valves, timers, tape recorders, programming devices, meteorological instruments—thousands of different instruments, and yet. . . .

Each complicated instrument consists of many not-too-complicated parts.

Let's see what some of these instruments do, and what they work on. The items listed in italics can be found in the Index; they have been discussed in this book.

DESCENT. The lander separates from its carrier ship, the Mars orbiter. The lander enters into a three-hour descent path. On the way it samples the atmosphere for charged particles—mainly *protons* and *electrons*—with an instrument using an *electric meter.* Farther down the descent path, a mass *spectrometer* measures the presence of uncharged (neutral) particles such as *neutrons* and *atoms.*

The mass spectrometer also tests for the presence of fifty of the elements every five seconds. The element nitrogen is of special interest—on earth it is present in all living things.

SLOWDOWN. Near the end of the descent to Mars, a parachute opens, slowing the lander by converting *mechanical energy* into pressure and *heat energy.* The lander's rate of descent and altitude are constantly measured by a landing *radar* and a *radar altimeter.* The lander is pulled by the gravitational field of Mars, whose surface force is about 38 percent as strong as Earth's.

LANDING. A moment after the parachute opens, three rocket engines are turned on.

The downward *thrust* of their exhaust gases converts *heat energy* into upward *mechanical energy*, to slow the final fall of the lander. The last three meters are free-fall, with the landing force taken up by three shock absorbers that convert *mechanical energy* into *heat energy* and *compression* stress.

CONTACT. Soon after the lander touches down, one of the two cameras aboard goes into action, to photograph the nearby soil and then the distant landscape. The cameras have *lens groups* of extremely high *resolving power*, almost totally free of *chromatic aberration*, working with *visible light* and *infrared light*.

The photographs are converted to electrical signals, *amplified* by *transistor* circuits and *telemetered* by *carrier waves* at *radio frequency* from an *antenna* to the orbiter. There they are further amplified and telemetered to Earth. All the instruments on the lander receive electric current from a thermoelectric generator that converts *nuclear energy* into electrical energy.

THE ATMOSPHERE OF MARS. Information about the atmosphere is gathered by several instruments. Weather information is obtained by a *thermopile* consisting of several *thermocouples*, a special type of *anemometer*, a *thermistor*, and a *barometer*. Chemical information is obtained by a *Gas Chromatograph/Mass Spectrometer*.

THE SOIL OF MARS. Eight days after landing, a surface sampler boom is extended ten feet out from its support. A *permanent magnet* at the far end picks up *ferrous* and other *magnetic* materials for examination by the cameras.

A small, mechanical shovel picks up soil and deposits it in a hopper. Some of the soil is moved into a test chamber to undergo *qualitative* and *quantitative* analysis. These tests are performed by *chromatographs* and *spectrometers*, one of which works by *X-rays*.

THE BIOLOGY OF MARS. The possibility of life on Mars is explored in two ways: visually and biochemically. Visually, the camera lenses can focus on anything the size of a grain of sand or larger. For smaller living things—*microorganisms*—three small biochemical laboratories go into action. Together, the three laboratories take up less than a cubic foot of space. Each one is set up to do a complete analytical process.

LABORATORY #1. "As I live and breathe" could be the motto for this laboratory. As *you* live and breathe, you exchange gases with the atmosphere: you take in oxygen and give back carbon dioxide. If living microorganisms are present in the soil sample, they, too, probably exchange gases with the laboratory atmosphere. The soil is steeped in a liquid *culture medium*, for microorganisms (hopefully) to feed on. If they are present they will "breathe," taking in one or more gases and emitting others. A gas *chromatograph* analyzes the atmosphere qualitatively and quantitatively, making periodic checks to see if there has been any change in the amount or kind of gases.

LABORATORY #2. Here the motto might be, "As I live and feed and breathe." As *you* live and *feed* yourself, you take in large amounts of carbon (in practically every food but salt). You use the carbon for your energy needs, combining it with oxygen. Eventually, you exhale it in the gas *carbon dioxide.*

Soil samples in Laboratory #2 are moistened with a nutrient containing *carbon-14,* which is *radioactive.* The soil is *incubated* for several days; then the atmosphere is tested for carbon-14. This is done with a C-14 detector, whose work is similar to a *scintillation counter* or a *Geiger-Muller counter.* The presence of carbon-14 in the atmosphere would indicate that something in the soil has taken in the nutrient and given off a gas containing carbon-14.

LABORATORY #3. Here the motto could be, "Animal, vegetable, or mineral?" On your planet there's no problem in telling the minerals (non-living) from the animals and vegetables (living). Nor is there usually any problem telling the animals from the vegetables. Vegetables (and all other green plants as well) can perform the process of *photosynthesis*: they take in carbon dioxide and water and combine them, with sunlight energy, to form a food compound—starch or sugar.

Does a similar process take place on Mars? A soil sample is placed under a sunlamp in Laboratory #3. The atmosphere is filled with *radioactive carbon dioxide* and other gases. After five days of "sunlight," the atmosphere is expelled. The soil is heated to a high temperature to expel any remaining carbon dioxide. Then the soil is tested for carbon-14. Its presence would indicate that something has taken carbon-14 out of the atmosphere and combined it with itself in the soil. In other words, something like photosynthesis could have taken place.

A Viking lander, with its three laboratories and many other instruments, is certainly a complicated package of complicated instruments. But complicated things are the sum total of simpler things. That's an encouraging thought as you look for the parts that make up the whole—as you find out more about. . . .

HOW SCIENTISTS FIND OUT

bibliography

Angrist, Stanley W. *Closing the Loop: The Story of Feedback.* New York: T. Y. Crowell Company, 1973.

Aylesworth, Thomas G. *The Alchemists: Magic into Science.* Reading, Mass.: Addison-Wesley, 1973.

———Mysteries from the Past: *Stories of Scientific Detection.* Garden City, N.Y.: Natural History Press, 1971.

Baldwin, Gordon C. *Inventors and Inventions of the Ancient World.* New York: Four Winds Press, 1973.

Bass, George F. *Archaeology Beneath the Sea.* New York: Walker & Company, 1975.

Branley, Franklin M. *Measure with Metric.* New York: T. Y. Crowell Company, 1975.

Briggs, Peter. *Laboratory at the Bottom of the World.* New York: David McKay Co., Inc., 1970.

Bronowski, J. *The Ascent of Man.* Boston: Little, Brown and Company, 1974.

Berger, Melvin. *Tools of Modern Biology.* New York: T. Y. Crowell Company, 1970.

———*Animal Hospital.* New York: The John Day Co., 1973.

Blaustein, Elliott, H. *Anti-Pollution Lab.* New York: Sentinel Books, 1972.

Dellow, E. L. *Methods of Science: An Introduction to Measuring and Testing for Laymen and Students.* New York: Universe Books, 1970.

Epstein, Sam, and Epstein, Beryl. *Michael Faraday: Apprentice to Science.* Champaign, Ill.: Garrard Publishing Company, 1971.

Ewbank, Constance. *Insect Zoo: How to Collect and Care for Insects.* New York: Walker & Company, 1973.

Fenten, D. X. *Ms.—M.D.* Philadelphia: The Westminster Press, 1973.

Freese, Arthur S. *Careers and Opportunities in the Medical Sciences.* New York: E. P. Dutton & Co., Inc., 1971.

Gallant, Roy A. *Explorers of the Atom.* Garden City, N.Y.: Doubleday & Co., Inc., 1974.

Goldstein, Philip and Metzner, Jerome. *Experiments with Microscopic Animals*. Garden City, N.Y.: Doubleday & Co., Inc., 1971.

Gregor, Arthur. *Bell Laboratories: Inside the World's Largest Communications Center*. New York: Charles Scribner's Sons, 1972.

Graham, Ada, and Graham, Frank, Jr. *The Careless Animal: Nine Ecological Detective Stories*. Garden City, N.Y.: Doubleday & Co., Inc., 1975.

Kay, Shirley. *Digging Into the Past*. Baltimore, Md.: Penguin Books, Inc., 1974.

Klein, Aaron E. *The Hidden Contributors: Black Scientists and Inventors in America*. Garden City, N.Y.: Doubleday & Co., Inc., 1971.

Lansing, Elizabeth. *The Sumerians: Inventors and Builders*. New York: McGraw-Hill, 1971.

Mac Nab, John. *The Education of a Doctor: My First Year on the Wards*. Simon & Schuster, Inc., 1971.

Magnusson, Magnus. *Introducing Archaeology*. New York: Henry Z. Walck, Inc., 1973.

Marks, Geoffrey, and Beatty, William, K. *Women in White*. New York: Charles Scribner's Sons, 1972.

Menard, H. W. *Anatomy of an Expedition*. New York: McGraw-Hill, 1969.

McKern, Sharon, S. and McKern, Thomas W. *Tracking Fossil Man: An Adventure in Evolution*. New York: Praeger Publishers, Inc., 1970.

Osen, Lynn, M. *Women in Mathematics*. Cambridge, Mass.: M.I.T. Press, 1975.

Polgreen, John, and Polgreen, Cathleen. *Backyard Safari*. Garden City, N.Y.: Doubleday & Co., Inc., 1971.

Pringle, Laurence, *Energy: Power for People*. New York: Macmillan, Inc., 1975.

——— (Ed.). *Discovering Nature Indoors*. Garden City, N.Y.: Natural History Press, 1970.

Russell, Helen, Ross. *Earth, the Great Recycler*. New York: Thomas Nelson, Inc., 1973.

Schwartz, George, I., and Schwartz, Bernice, S. *Life in a Log*. Garden City, N.Y.: Doubleday & Co., Inc., 1972.

Schwartz, Julius. *It's Fun to Know Why: Experiments With Things Around Us.* Second Edition. New York: McGraw-Hill, 1973.

Shannon, Terry, and Payzant, Charles. *Antarctic Challenge: Probing the Mysteries of the White Continent.* Chicago, Ill.: Childrens Press, 1973.

Silverberg, Robert. *Clocks for the Ages: How Scientists Date the Past.* New York: Macmillan, Inc., 1971.

Simon, Seymour. *Science at Work: Projects in Oceanography.* New York: Franklin Watts, Inc., 1972.

Trowbridge, Leslie, W. *Experiments in Meteorology.* Garden City, N.Y.: Doubleday, Inc., 1973.

Sootin, Harry. *Easy Experiments with Water Pollution.* New York: Four Winds Press, 1974.

Williams, Patricia, M. *Museum of Natural History and the People Who Work in Them.* New York: St. Martin's Press, Inc., 1973.

Wilson, Mitchell. *Passion to Know.* Garden City, N.Y.: Doubleday, Inc., 1972.

Webster, David. *Track Watching.* New York: Franklin Watts, Inc., 1972.

Wong, Herbert, H. and Vessel, Matthew, F. *Pond Life: Watching Animals Find Food.* Reading, Mass.: Addison-Wesley Publishing Co., Inc., 1970.

index